# BORN TO FLY

# BORN TO FLY

## The Untold Story of the Downed American Reconnaissance Plane

SHANE OSBORN

with Malcolm McConnell

BROADWAY BOOKS

New York

Broadway Books titles may be purchased for business or
promotional use or for special sales. For information, please write to:
Special Markets Department, Random House, Inc.,
1540 Broadway, New York, NY 10036.

PRINTED IN THE UNITED STATES OF AMERICA



Credits for photo insert: pp. 1–4 Courtesy of Shane Osborn; p. 5 U.S. Navy Photo—
Courtesy of Shane Osborn; U.S. Navy Photo—Courtesy of the Lockheed Martin
Aeronautics Company; p. 6 U.S. Navy Photos—Courtesy of the Lockheed Martin
Aeronautics Company; p. 7 U.S. Air Force photo by Tech. Sgt. Paul Holcomb—
Courtesy of Shane Osborn; photo of President George W. Bush and Shane Osborn—
Courtesy of Shane Osborn; p. 8 U. S. Navy photo by Photographer's Mate 2nd Class
Bob Houlihan—Courtesy of the White House.

Visit our website at www.broadwaybooks.com

FIRST EDITION PUBLISHED 2001

Cataloging-in-Publication Data is on file with the Library of Congress

ISBN 0-7679-1111-3

Designed by Erin L. Matherne and Tina Thompson

1 3 5 7 9 10 8 6 4 2

# Contents

# BORN TO FLY

Chapter One

# KADENA AIR BASE, OKINAWA

## 1 April 2001, 0430 Hours

I stood on the hardstand, staring up at the big gray-and-white airplane. The wings and four turboprop engines of the EP-3E ARIES II reconnaissance aircraft were outlined sharply in the portable floodlights the maintenance people had used in the night to prepare for today's mission. But above the lights, the cold predawn darkness was full of stars. After several days of thunderstorms, during which we'd had to scrub one mission, it looked like we were going to have a perfect day to fly.

By the traditions of Naval Aviation, the final walk-around inspection of the plane was the responsibility of the Aircraft Commander. I took this task very seriously, even though third pilot Lieutenant Junior Grade Jeffrey Vignery and flight engineer Senior Chief Nick Mellos had already conducted their own rigorous walk-around inspections while I had been getting my pre-mission briefing from the intelligence people supporting Fleet Air Reconnaissance Squadron VQ-1.

As I ran my hand across the fiberglass skin of the forward weather radome in the nose, Senior Chief Mellos emerged from the shadows. He'd been smoking an old sailor's "hot butt" safely away from the fully fueled aircraft.

"Hey, Senior," I kidded him, "getting in that last precious hit of nicotine?"

"Roger that, sir," he said, returning my grin. "Never know where the next one's coming from."

With his shaved head and graying mustache, Senior Mellos looked like a middle-aged pirate. But his green Navy flight suit covered muscular shoulders and a barrel chest. He'd been in the Navy for twenty-eight years, most of that time flying on one variant or another of the P-3 Orion maritime patrol plane. That was two more years than I'd been alive. Back at VQ-1's home base at Naval Air Station, Whidbey Island, Washington, Senior Mellos and I rode our Harleys on off-duty hours. On overseas deployments, we partied with the rest of the crew. But on missions, our relationship was strictly professional. It was also based on mutual respect. I relied heavily on Senior Mellos's long experience as a flight engineer to keep me up to speed on the condition of the complex airplane and to help keep all of us out of trouble during an airborne emergency. So when he pronounced this aircraft "a solid airplane," with only minor maintenance problems, I felt certain we could make our takeoff time.

We flew our planes hard in VQ-1. Even though we were based along Puget Sound, our operational work was overseas. At any given time, the squadron had planes and crews detached—on "Dets"—in the Western Pacific, in Bahrain flying missions along the Arabian Gulf or over Kuwait for Operation Southern Watch, which enforced the no-fly zone in Iraq, or conducting counternarcotic surveillance in South America. I had flown from all those sites since joining VQ-1 in April 1999. But this Det to the Far East, which had begun in early March, was my first as a mission commander, a job that required the flying skills of an Electronic Warfare Aircraft Commander (EWAC) with the in-depth knowledge of the squadron's complex surveillance mission.

Using my flashlight, I inspected the nose wheel for tire wear and the shock-absorbing gear strut for leaking hydraulic fluid. I walked under the left wing and repeated the process with the main landing gear on that side, then peered up inside the dark wheel well to check the pressures in the two engine fire extinguisher bottles. Next, I stepped forward and reached up to grab the red-striped tips of the heavy gray alloy propeller blades of the number one and two engines, checking for obvious nicks or wobble. The four big Allison engines could each produce 4,600 shaft horsepower. It was essential the props were smooth and perfectly balanced or their runaway vibration could damage the aircraft.

The Big Look radar dome, a bloated gray doughnut beneath the plane's forward belly that we called the "M&M," and a narrower canoe-shaped antenna cover farther aft, were both solidly attached. Smaller antenna mounts beneath the wings and near the tail were also intact and undamaged. All told, the EP-3E bulged and bristled with sensor pods, disrupting the original airframe's streamlined exterior and producing a lumpy appearance that some aviators found ugly. But I've never seen a bad-looking airplane.

This array of external equipment revealed the plane and the squadron's mission: The acronym ARIES II stood for Airborne Reconnaissance Integrated Electronic Systems II. EP-3Es were one of America's most capable platforms for collecting signals intelligence (SIGINT). With our sensitive receivers and antennas we could look over the horizon from international airspace in support of the ships and aircraft of the Fleet to pinpoint a wide range of radar and radio emissions. This was required to develop an accurate picture of what's been called the "electronic order of battle," which might include the electromagnetic activity of surface-to-air or surface-to-surface missile systems. We also were capable of providing direct "real time" tactical electronic reconnaissance during combat operations to our fighters and strike aircraft so that they could better avoid threats and locate targets.

To conduct this mission successfully, the EP-3E had a crew of twenty-four, the largest of any U.S. military aircraft. Because our missions routinely lasted over ten hours, we carried three pilots and two flight engineers. On this Det, I was both EWAC and Mission Commander. The second pilot, or 2-P, was Lieutenant Patrick Honeck, with Jeff Vignery, the 3-P, flying his first Det to the Western Pacific. Senior Chief Nick Mellos, with over 8,000 flying hours, was one of the most veteran flight engineers in the Navy, but the other FE, Petty Officer Second Class Wendy Westbrook, was a cool hand and a fast study. I knew from experience I could trust her in the seat between the two pilots. Our navigator was Lieutenant Junior Grade Regina Kauffman, who had not yet officially qualified for her position.

The "guys in back" were the electronic warfare operators (EWOPs), cryptologic technicians, reconnaissance equipment specialists, and the special operators supervised by special evaluator Lieutenant Marcia Sonon. Lieutenant Junior Grade Johnny Comerford, the senior evaluator (SEVAL), had overall responsibility for these reconnaissance personnel. In total, eighteen people worked at in-line computerized consoles set along the sides of the cabin, the long, narrow "tube" that was broken up by the head (toilet), the small galley and its cramped booth that looked like something out of a 1950s diner, and a stacked pair of curtained bunks for the off-duty pilot and FE who could grab some sleep on long missions.

Adequate rest was actually a critical safety factor for our flight crews. Once the reconnaissance crew had checked out their sensors after takeoff, they could lean back in their seats and nod off on the long, slow cruise to our reconnaissance track. But even on autopilot, two pilots and one FE had to stay awake and alert for hours on end because we flew Visual Flight Rules—Due Regard, which meant that unlike civilian airliners, we had to rely on ourselves, not on ground controllers, for our safe separation from other aircraft.

I climbed the folding ladder to the main cabin entrance on the portside aft, and Senior Mellos secured the hatch. Now I was engulfed in the familiar scent of the airplane: warm electronics, a slight whiff of jet exhaust from the Auxiliary Power Unit in the nose, and hot coffee in the insulated mugs some of the crew had carried on board. A few of the people looked sleepy. Today had been a "three-for-five," an 0300 briefing for an 0500 takeoff, which meant most of us had gotten up at two, rushed through the shower, and managed to grab a quick breakfast in our rooms before heading off for separate briefings depending on our specialties.

Now we had all reassembled in the cabin so that I could give the final "planeside" brief. Even though we'd all been through this together as a crew a dozen times on this Det alone, I again reviewed procedures for ground and air emergencies, which might include the need to bail out or ditch the plane at sea. On entering the cabin, the crew members had taken a grease pencil and printed their names on their position line of the plastic ditching placard on the head door. Each line marked the number and storage rack of that position's parachute. Also stowed aboard the aircraft were our SV-2s, a one-piece combination survival vest–Life Preserver Unit (LPU), and custom-fitted flight helmets. We all carried our own Nomex fire-resistant aviator gloves.

I completed briefing the ditching procedures by pointing to the two big life rafts stowed in the orange rubber soft packs near each over-wing emergency exit hatch. Now Johnny Comerford, the SEVAL, briefed the crew on bailout procedures. He was the jump master who would supervise the distribution of parachutes, make sure everyone was lined up correctly, holding the overhead grab rail, and then see them out the hatch of the main cabin door at one-second intervals.

We could all practically recite these instructions by rote, but this was a U.S. Navy aircraft on a demanding mission, and we followed the book in everything we did. In our case, that book was the Naval Air Training and Operational Procedures Standardization (NATOPS). Everything the flight crew did aboard the EP-3E was covered in that black three-hole binder thick with checklists, schematics of instrument layouts, and wiring diagrams.

After Johnny had reviewed bailout procedure, I briefed the crew on the weather. "It looks like good weather en route to the track orbit and back to Kadena. Excellent visibility and no reported turbulence at our assigned altitudes. Briefed mission time is just over nine hours today."

We were headed down the coast of Asia to the South China Sea. Once on track, we would fly our reconnaissance track in international air space south of China's Hainan Island and north of the Philippines. Our squadron had been flying such missions in this area in one kind of aircraft or another for decades. Signals intelligence planes from a number of countries, including China, fly similar reconnaissance missions. With the increased military dependence on sophisticated radars and data links, airborne SIGINT forms a vital part of a military commander's resources. So we were not overly apprehensive about the mission, even though someone had joked that it was April Fool's Day.

But we did expect that Chinese fighters would intercept us in international air space at some point along our track. Once Chinese radar picked us up, they would usually dispatch a pair of twinjet F-8 II "Finback" interceptors from the People's Liberation Army/Navy (PLAN) to look us over. This had been going on for years. Recently, however, these intercepts had become increasingly aggressive and had reached the point of being downright hazardous to both the fighters and the EP-3Es as the Chinese flew closer and closer to our planes. On January 24, for example, a Chinese pilot came within thirty feet of an EP-3E, hanging just off the American plane's left wing before he firewalled his throttles and buffeted the EP-3E with jet wash in a crazy, dangerous maneuver called "thumping."

This was reckless airmanship. Chinese pilots were not well trained in formation flying. In fact, when they intercepted us, they always approached in loose trail because they didn't trust their flying skills enough to maneuver close to one another.

Although originally designed in the 1960s, the F-8 II Finback modification was a high-performance, air-superiority fighter capable of flying Mach 2.2 at almost 60,000 feet. The EP-3E flew much lower and slower, usually cruising at around 180 knots at an altitude between 22,000 and 24,000 feet. In order for a F-8 II to maneuver close alongside an EP-3E at that speed and altitude, the Chinese fighter had to fly "dirty," with flaps and slats extended from his thin delta wings, nose tilted awkwardly upward in what we call a high Angle of Attack (AOA). Then the Chinese plane is really struggling to maintain lift with the pilot delicately balancing his control stick and engine thrust to keep him in the air. But his plane is barely stable. Jet engines require a long spool-up time to provide adequate thrust, and at these speeds, the pilot of a nose-high Finback needs a much greater throw of his stick to control the fighter's movements.

That's called "mushy" flying.

The term is a good description of how Chinese fighters had been performing in their aggressive intercepts in recent months. It was one thing for them to corkscrew through the sky a quarter mile off an EP-3E's wing, trying to keep pace with a reconnaissance plane lumbering along on autopilot. It was something else entirely for them to bob and wobble their way along within fifty feet—or less—of us. One of their pilots had even flown close enough to another American reconnaissance plane to hold up a sheet of paper with his e-mail address to be read. These dangerous intercepts had prompted the U. S. government to raise a formal protest with the People's Republic of China during the winter. But it looked like the Chinese Navy pilots flying the F-8s had not gotten the message because we'd been intercepted even closer than that on this Det.

I finished my brief and made my way forward up the narrow center aisle toward the flight station. On the port side, Johnny Comerford's EWOPs were getting settled into the consoles. Cryptologic Technician Jeremy Crandall was stowing one of the binders of classified material Johnny had brought on board. Others along the line were plugging in their headset cords to the Intercom Communications System (ICS) for voice checks before takeoff. The ICS allowed the crew to speak comfortably above the whining roar of the big engines. By selecting different "loops" in the system, the Evaluators, EWOPs, and technicians could speak directly with each other, or Johnny could contact the flight station without disturbing the rest of the crew. But there was a general public address system with speakers powerful enough for all of us to hear important announcements over the engines.

At the secure communications station facing forward at the front of the cabin, Lieutenant Junior Grade Dick Payne was entering codes into the computerized radio equipment that allowed us to contact our chain of command directly with encrypted messages.

Slightly forward on the starboard side, our navigator, Lieutenant Junior Grade Regina Kauffman, was hunched over the narrow chart table, plotting the first leg of our route southwest from Okinawa. Regina is not a tall woman; she probably stands five feet in her steel-toed flying boots. Passing the Nav station, I had the impression she'd been swallowed by her banks of instruments and was hidden inside her flight suit and the green nylon mesh and floppy rubber horse collar of her SV-2.

Senior Mellos was in the central flight engineer's seat and Jeff Vignery, the 3-P, in the copilot's seat on the right. Like the other pilots and naval flight officers in our squadron, Jeff had acquired a personal call sign. In his case, it was "Jefe" (Spanish for "boss"), picked up during his first Det to South America. Jeff was a devoutly religious guy from Kansas with short carrot-colored hair. Relatively new to the EP-3E, he was a good pilot who worked hard to learn our demanding profession. I had tagged Johnny Comerford "Johnny Ballgame" from the 1980s movie *Johnny Be Good* because I could always count on him to hit the town with me on weekends and have a good time.

Regina Kauffman was "Reggie." And Pat Honeck was "Meeso" from the R-rated song lyric "Me so horny" because he came to the squadron already tagged with that call sign. I was either "Sugar Shane" Osborn from my naturally sweet disposition (ha!) or "Angry" Osborn when I worked my squadron ground job as a crew scheduler in Operations back at Whidbey Island. I was known to have blown up at people once or twice when they tried to find excuses not to go on a Det. After all, we were Naval Aviators flying operations, and those operations took us around the world. And that took us away from our loved ones. Sailors had always accepted this. You couldn't talk the talk unless you were willing to walk the walk.

"What's up, Jefe?" I asked Jeff as I slid into my seat and buckled my harness.

"Senior says he's got us a strokin' airplane," Jeff replied, confirming Senior Mellos's initial evaluation that any problems in the plane's complex maze of electronic, electrical, mechanical, and hydraulic systems were practically nonexistent. All the banks of switches and instruments on the flight deck gleamed. Taking advantage of the bad weather earlier in the week, our crew had cleaned the entire airplane inside and out, vacuuming, sweeping, and polishing. Even though the airframe had been built as a sub-hunting P-3 Orion maritime patrol plane in 1969, and there were the inevitable dents, chips, and scuffs visible throughout the cabin, this aircraft had always received excellent maintenance and had been completely overhauled in its conversion to an EP-3E. The process sometimes reminded me of the old Nebraska farm joke I'd heard as a kid: "This is my grandpa's ax. My dad changed the head, and I changed the handle."

Speaking through our headset mikes, we began the familiar challenge-and-response chant of the flight station checklists leading through Engine Start to final preparation for takeoff.

"Parking brake," Jefe intoned.

I reached around the control yoke column to tap the horizontal bar with my gloved right hand. "Set."

We patiently worked down the long checklist. One item was verifying that the two pilots' circular gyro artificial horizons were running. This instrument is divided between a black bottom representing the earth and a white upper hemisphere for the sky. The gyro displays our angle of turn and bank and the degree of climb and dive. The horizontal situation indicator (HSI) beside each pilot's gyro was an electronic compass that could be tuned to radio navigation beacons.

"Flaps," Jeff called, touching the sliding blue wing-shaped lever in the flap panel on the right side of the console. "Set for takeoff."

I leaned toward the left windscreen and signaled the lineman below to remove the wheel chocks. Senior Mellos handed me a clipboard with the fuel quantity totals, center of gravity, and calculated takeoff speed. This was vital information. Today we carried 58,000 pounds—29 tons—of jet fuel. That would leave us the required minimum of 8,000 pounds "on top" when we returned back to Kadena after the nine-hour mission. If the field were closed for weather or any other reason, this reserve would guarantee us two hours to continue on to an alternate airfield in Japan.

The Kadena runway was dry, and the cool air dense, which made for optimum lift on takeoff. This was a good thing because our total weight was 140,000 pounds, only 2,000 pounds under the maximum allowable gross weight, and we were going to need a long takeoff run.

During the Engine Start sequence, Senior Mellos reached up to the overhead instrument panel and turned the selector switch left to number two, the second engine in from the end of the left wing. I advanced the number two power lever on the left console to the Start position. "Start number two," I said.

"Starting number two," Senior Mellos replied, stabbing the red button with his thick finger.

The turbine whined and the prop began to turn, slowly at first, then with increasing speed as the RPM built up. Jeff and I had a count running; if the engine didn't start within sixty seconds, we'd have to shut it down. But the engine instrument panel showed the shaft horsepower, turbine inlet temperature (TIT), prop RPM, and fuel flow climbing steadily. Even with only one engine running, the howling rumble on the flight station made normal speech impossible. That's why we always used our ICS headsets.

Senior Mellos disengaged the start button and switched the selector to number three, the inner engine on the right wing. We always started numbers two and three first because they had engine-driven compressors from which we ran our air-conditioning, heating, and cabin pressurization. Now we started numbers one and four and carefully monitored the vertical bank of instruments for each engine on the central panel, making sure their performance was normal when we throttled back to Ground Idle.

While Senior Mellos and Jeff finished up the final pre-takeoff items on the checklist, I again studied the form with the fuel quantity, center of gravity, airfield temperature, and barometric pressure that comprised our takeoff speed information. Then I turned my ICS switch to speak to the crew in the cabin. "Set Condition Five," I ordered. Now everyone would turn and lock their seats facing aft and make sure their harnesses were snugged down for takeoff. A minute later, Johnny replied, "Condition Five set."

After a final check of the instruments, I gave my takeoff briefing to Jeff and Senior Mellos. On an EP-3E, engine performance is measured by turbine inlet temperature and shaft horsepower. The props are attached to a geared hub and automatically maintain 100 percent RPM while the pitch varies with turbine speed. Because we were so heavy today, we would need maximum continuous power.

"Okay, Senior," I recited in a quick, familiar sequence, "set 1010 TIT on my command. Jeff, call out eighty knots for power and airspeed checks. We're looking for 4,600 shaft horsepower and we'll abort with less than 4,400 shaft on all four engines. Call out any malfunctions by type and engine number. We've got a good long runway, so there's no refusal today." By this I meant there was not a point before reaching rotation speed at which we could no longer safely abort without running out of pavement if the engines weren't performing. "Call out rotate at 131 knots. If we have a malfunction prior to rotate, I'll abort the takeoff. If we have a prop malfunction, I'll call for the appropriate E Handle prior to bringing the power lever to Flight Idle."

The four yellow-and-black striped engine Emergency Handles occupied a prominent position at the top center of the main cockpit console. Pulling an E Handle automatically shut down an engine both mechanically and electrically. Then, the four propeller blades of the powerless engine would feather, locking in position so that their sharp edges faced directly into the airstream to minimize drag.

"If we have a malfunction after rotate at 131 knots," I went on, intentionally repeating the rotation speed for emphasis, "then we're going to continue our assigned climb out and handle the malfunction once we're safely airborne in accordance with the NATOPS. Any questions?"

They had none. After a final round of checklist challenge-and-response items, Jeff and I turned our control yokes right and left, peering out our side windscreens to make sure the long ailerons on the wings' trailing edges rose and dipped properly. I pushed my control yoke forward and hauled back, and the lineman standing beneath my window signaled the elevators in the tail's horizontal stabilizers moved freely. We were ready for taxi. The lineman snapped off a sharp Navy salute, which I returned.

Jeff called the Kadena tower requesting permission to taxi to Runway 4 Left.

"Kilo Romeo 919," the tower replied with our mission call sign, "clear to taxi."

Now the lineman was well clear of the left wing and the slashing propellers, signaling with lighted orange-cone taxi guides. I released the parking brake and eased the four power levers forward with my cupped right hand gripping their knobs. Given our weight today, it took some engine thrust to overcome inertia and start us moving. With my left hand I held the small alloy circle of the nose wheel steering system. We trundled down the wide taxi ramp, then made a turn. Swinging through the next turn on the taxi ramp, Jeff and I verified that the needles of our HSI compasses tracked from northwest toward north. We were almost at the end of the long ground checklist. Completing it day in, day out on either training or operational flights or in the simulator back at Whidbey was just one of the disciplines we accepted as Naval Aviators. You never took shortcuts in peacetime. Ever.

"IFF," Jeff said, referring to the radar transponder that squawked our aircraft identification for civilian ground controllers. We would switch this system from Active to Standby once we left Japanese air space. "Set."

"Flaps," he said, repeating this vital checklist item. "Flaps set for takeoff." We both glanced down at the blue flap lever.

I eased the power levers back slightly, and the props whistled with the decreased pitch. Even though the taxi ramp was smooth, the big plane swayed, and the nose dipped with the change in speed. By then my eyes were well adjusted to the moonless night outside, so I dimmed the instrument panel a bit more as we reached the end of the runway. One good thing about such an early takeoff was that the normal training day for the Air Force and Marine Corps planes at the big base had not started.

That was fine with me, because I wanted plenty of room. This heavy, if I lost an engine while rolling above a hundred knots, the plane would not be very controllable and I'd have to act fast to overcome a potentially serious swerving problem. Although we routinely took off at this weight in EP-3Es, it was definitely pushing the outside of the envelope.

I was certainly attentive, but not tense. A few weeks earlier up at our Det site in Misawa on the northern end of Japan's Honshu Island, we had taken off just as heavy during a six-inch snowfall in minimum visibility and ceiling. In fact, I hadn't been able to see the distance-marking boards on the sides of the runway, just a blur of lights. And the nose wheel steering had been useless on the icy taxi ramp, so I'd had to use differential engine power to maneuver. As soon as I had rotated and had called "gear up," we were in the clouds, flying on instruments, watching intently for any sign of icing on our props and wings.

"Harnesses," Jeff said, sounding off the last item on the checklist.

"Set," all three of us repeated, making sure the buckles were securely locked.

"Kadena tower," Jeff called, "Kilo Romeo 919 ready for takeoff."

"Kilo Romeo 919," the tower answered, "takeoff four Left. Wind 010 at eight. Cleared for takeoff."

We had eight knots of wind, almost right down the pipe, which would give us a slight airspeed boost.

I moved the power levers forward again and tapped the top of the rudder pedals very lightly with my toes to test the brakes one last time. Even with this slight pressure, all three of us slid forward against our harnesses. The plane had very strong brakes, and they were tricky. You had to steer with the rudders on the takeoff roll, but if your toe accidentally touched a brake, you would blow a tire in a second.

Then you had to immediately pump the opposite rudder to overcome the swerve. If you were really unlucky and blew both tires on a main gear, you might not be able to keep the plane on the runway. You didn't want to dwell too long on what might happen if the plane spun off the runway at 100 knots, a landing gear collapsed and the wing scraped along the pavement, bending props and spewing up sparks. We carried most of our maximum 58,000-pound fuel load in wing tanks, close to the engines' hot turbines.

But a qualified pilot had to be able to safely abort the takeoff after a blown tire. That required slowing the plane with reverse propeller pitch, a maneuver we performed by pulling back and lifting the power levers into "ground range" and adding thrust, so that the flattened prop pitch created drag. And this, too, could be tricky when you already had the terrible drag of blown tires pulling you to one side.

I had never suffered a blown tire on takeoff. But I had *flown* this malfunction a score of times in simulators. It was an experience that always left me sweating.

So, we learned early to respect our brakes. From their first days in flight station mock-ups, P-3 pilots were taught to steer with the balls of their feet on the rudder pedals, the heels of their flight boots resting on the deck.

I scanned the instruments again. "Everybody ready to go?"

"Roger," Senior and Jeff both replied.

Positioning the plane on the runway centerline, I slid the power levers forward. "Set 1010, Senior."

Senior Mellos continued to push the power levers full forward, as I placed my feet on the rudder pedals and guided us down the centerline with the nose wheel steering. The plane gathered speed fast. My eyes jumped back and forth from the centerline to the lighted marker boards at the sides of the runway to the airspeed indicator on the left of my instrument panel. At around 50 knots, I felt the rudder come alive through the soles of my boots, and took my hand off the nose wheel steering. Now I was holding the yoke forward to keep the nose down and prevent bouncing as we accelerated. Pumping my feet smoothly, I kept us centered on the runway.

About ten seconds after I had begun steering with the rudder, Jeff called out, "Eighty knots."

We both scanned our airspeed indicators to make sure there were no splits, that we were registering the same. Mine indicated 80 knots and rising. The boards were flashing by now. We also scanned our engine gauges for malfunctions. There were none. The engines roared with solid power.

"Rotate," Jeff said.

Before pulling back gently on the yoke, I again quickly scanned the TIT, shaft horsepower, and airspeed. We were passing 133 knots. I eased back on the yoke and slowly pulled the plane off the deck and let it fly itself with a slight nose-up angle. This heavy, the last thing you wanted to do was overrotate. If you did with this fuel load, the plane would sink back to the runway. At the very least, you'd scrape the hell out of the tail and ruin several million dollars worth of high-tech antennas. But sinking like that could also be a lot worse. My eye shot to the small Vertical Speed Indicator (VSI) to the right of the gyro. The arrow was sliding up.

"Positive rate of climb," I said. "Gear up."

Jeff's hand was already reaching for the tire-shaped end knob of the lever. "Gear up," he replied, adding a moment later, "and locked." The nose and main gear came rumbling up and seated in their wells with a comforting thump.

Climbing clear of the runway now, I pushed in a lot of right rudder to compensate for the P-Factor of the props whose combined torque would yaw us hard left unless I did so. As Jeff got our departure heading from the tower, I watched the airspeed closely. Even this heavy, we were quickly approaching 160 knots. The maximum speed for Takeoff Flaps is 190. But I didn't want to retract them to the Maneuver position a moment too soon because we needed every bit of extra lift they gave us. I also did not want to overspeed the flaps and damage them.

Just above180 knots, I told Jeff, "Flaps to Maneuver."

Senior Mellos reported the cabin was pressurizing normally as we climbed and the prop synchronization was on the money. Approaching 200 knots, I called to Jeff to retract the flaps.

"Speed checks," he read from his checklist, "flaps coming up."

Again, both our airspeed indicators matched and the flaps had retracted completely. It looked like there were no showstoppers today. Climbing through 6,000 feet, we left the long, narrow island of Okinawa and rose into the dome of stars above the dark Pacific.

The quarter moon had set at midnight, so these stars had no competition in the night sky. Within minutes, however, a faint plum-colored band appeared on the eastern horizon, false dawn. One fat, unblinking star hung above the brightening band, probably Venus or Mercury. Regina would know, but she was stuck just behind us at the Nav station, busy monitoring the global positioning system (GPS) satellite receiver that was coupled to our two redundant inertial navigators as she followed our course south-southwest.

Japanese Air Traffic Control cleared us for a climb to a cruise altitude of 18,000 feet in slow, 2,000-foot increments. This gave us time to burn off some fuel. Before the people in back could undo their harnesses and swivel their seats back around, a crewman got up and went to a side window to check the engines for smoke and inspect the nooks and crannies where hydraulic lines and pumps were hidden, looking for leaks. There were no problems to report.

"Set Conditions Two, Three, and Four," I called over the ICS. Now the crew could finish preparing their equipment or just relax as we droned south into the early spring sky. Although I had flown hundreds of takeoffs during my five years as a Navy pilot, I still found this demanding, disciplined process completely satisfying. April Fool's Day or not, it *was* a great morning to fly.

But we still had work to do on the flight station before we could engage the autopilot for our cruise toward the trackline. Jeff and I carefully scanned the engine instruments, watching for any sign of overheating or power loss during the long climb. Shaft horsepower, TIT, prop RPM, and fuel flow were normal for each engine. Senior Mellos flipped switches on the overhead panel, fine-tuning propeller synchronization to minimize vibration. We were working as a close-knit team, with no additional need for checklists.

By 0625 Okinawa time, almost ninety minutes after takeoff, we were at 21,500 feet, heading toward the Luzon Strait between the Philippines and Taiwan. Because the EP-3E had so many external antennas, our airspeed was restricted to 250 knots unless there was an emergency. Today we droned along on autopilot at the usual 210 during our cruise to the trackline. Once on track, we would slow down to 180 knots to conserve fuel. As forecast, the morning was beautifully clear, with just a few broken clouds far below. Petty Officer Wendy Westbrook came up to swap out with Senior Mellos in the FE's seat. Even though he had downed one of his big mugs of notoriously strong coffee, I guessed he was probably now snoring away in his rack back aft.

After fifteen minutes of chatting with Jeff and Wendy, I decided that sleep sounded pretty good. Jeff could fly the airplane from the left seat and Pat Honeck could sit in the right. It would be good for Jeff, Pat, and Wendy to have complete charge of the flight station for a while. When we got on track, the pilots and flight engineers could all swap back and forth more frequently, but I would take the first break now because I needed some rest.

Out of habit, I held the overhead grab rail as I made my way aft through the cabin, even though there was no turbulence. Some of the crew were still loading software in their equipment. Others sat at their positions, heads down on their folded arms like kids sleeping at school desks. As usual, a few people had gathered in the galley booth to thumb through training manuals and magazines.

Stretching out in the upper rack, sleep did not come quickly. One of my shortcomings as a reconnaissance pilot was that I couldn't just flop down and grind out a few *zzzs*, then wake up refreshed fifteen minutes later the way so many other guys could. I've always been a light sleeper, and an EP-3E rumbling along toward its track was not exactly a restful environment. After I pulled the curtain tight around the bunk to shield my eyes from the high-altitude sunlight, I scrunched up the pillow and reseated my foam rubber earplugs. But half my mind was still listening for any change in the engine noise and bracing for an unanticipated bank or turn. I'd been this way since the beginning of the Det, probably because I hadn't gotten used to the responsibility of being a mission commander.

But becoming a pilot had been my ambition for as long as I can remember. In fact, flying formed my first vivid memory.

My mom, Diana, and my dad, Doug "Ozzie" Osborn, had a home in the tiny village of Loomis, a collection of a few houses north of the small city of Mitchell in eastern South Dakota. At any given time we might have had forty neighbors. My folks ran several businesses in Mitchell, including an upholstery shop, a bookstore, and a pizza parlor right across the street from the Corn Palace, one of America's original roadside attractions, a multicolor Taj Mahal of domes and minarets decorated each year with thousands of ears of corn.

But my older sister, Lynnette, and I didn't see too much of the bright lights of Mitchell. She attended a one-room school in Loomis, and I just stayed home as a little guy. Although Mom was a registered nurse, she and Dad devoted all their time to their small businesses. And the long hours they worked paid off.

Dad did find time to take me flying with our neighbor, a sheep farmer named Lyle Brewer, who flew his bright yellow Piper J-3 Cub from a pasture near our house.

Lyle, who had been a pilot his whole adult life, had bought the little two-place tail-dragger from a military surplus catalogue. The former liaison plane came in several pieces stowed inside crates, and Lyle carefully reassembled the aircraft from the accompanying instruction manual. Although I don't remember that part very well, my folks say I used to sneak out of the house and watch Lyle working patiently in his barn. I would have been almost four that spring.

Lyle must have done a good job, because the FAA inspector from Mitchell signed the airworthiness certificate. I took my first flight with Lyle and Dad almost as soon as the government man had signed off that the plane was safe to fly.

*That* experience I definitely do remember. I sat strapped in on Dad's lap in the narrow rear cockpit. Lyle hit the starter; there was a chug of smoke from the engine before it caught. Then the wooden prop spun to life. He opened the throttle, and the whole plane began to tremble and shake as we bounced down the pasture. The tail wheel lifted free of the grass just as we reached a little dip, and we gathered airspeed on the down slope. A few seconds later, the ground fell away and we were climbing.

"How do you like it, Shane?" Dad yelled over the roar of the motor.

"We're flying!" I shouted back, grinning like it was Christmas morning.

When Lyle got us up to a few hundred feet, he waved back, and my dad loosened the tight seat belt. The Piper Cub has hinged windows that fold down and snap open. Dad freed the right and left windows in our cockpit, and I immediately leaned over to get a better view.

"Whoa," he said, keeping his arm around my chest.

I knew Dad was there protecting me, but the view of the earth passing slowly below captured my senses. Square fields of green alfalfa or tan stubble drifted by. The rows of tasseled corn that always seemed so tall on the ground looked like grass. Twisting ribbons of cottonwood trees marked the hidden streambeds. As I leaned out the window, the warm propwash whistling around the wing strut stroked my face. We bucked in a sudden thermal current. I could smell the ripening crops and clapped with pleasure.

Lyle made a banking turn to the south, added more throttle, and began to climb again. Thrusting my arm into the slipstream, I flattened my hand like a little airplane and pretended to be flying, my body an extension of the Piper Cub.

Then I hunched forward to peer over Lyle's shoulder, now just as fascinated to see what he was doing to fly the plane as I had been by the view. He had one hand on the throttle knob, the other on the control stick between his lanky knees. His scuffed boots rested on a pair of rudder pedals under the simple instrument panel.

Suddenly I recognized the second set of controls in our tandem cockpit. The stick and rudder pedals were moving in unison with Lyle's actions. I reached out and gently touched the stick just as Lyle banked steeply right toward Lake Mitchell. The wings seemed to stay level while the green prairie horizon tilted with the bank. *This* was how you flew an airplane: Pull the stick back, and the nose went up. Move the stick in one direction and the plane banked that way. I hadn't yet figured out the role of the throttle.

"I want to fly," I told my dad.

"Not today, Shane."

Lyle took us over the lake. There were miniature water skiers down there, carving white curves in the water. Before this flight, I'd always thought of the water ski boats with their howling motors as the fastest, most powerful machines in the world. But from up here they looked like toys.

With the sun getting low, Lyle banked back northwest. I watched the left wing dip, the right rise, and the fields tilt beneath us. Then we passed over the familiar cluster of white houses and peaked roofs of Loomis. Except there was nothing familiar about our little town at all. It looked like a Monopoly board with telephone wires. I leaned farther out, eager to spot our house.

"Come on, Shane," Dad said, pulling me back onto his lap and tightening the seat belt, "time to get ready for landing."

Strapped back down, I couldn't see over Lyle's shoulder. But the engine noise grew softer. Dad reached over and closed the side windows, and our cockpit seemed almost silent. Drifting down through the afternoon sunlight, I felt a deep sense of satisfaction. This was better than any carnival loop-the-loop ride.

Lyle suddenly gunned the throttle as he leveled off to fly across the pasture and scare away any sheep that might have gotten loose from the pen to graze out there. The grass was empty. Once more the stick in our cockpit turned, then moved back to the center. The engine was only a whisper. My view ahead was blocked by Lyle's seat, but I could look out to either side and see the pasture rising toward us. Takeoff had been a thrill, but landing was magic. It was as if we were pulling the whole huge earth into the sky.

The left wheel kissed the grass with a slight jolt, then the right. Our cockpit sank back as the tail wheel touched the pasture. At this slow speed, we didn't roll out more than 300 feet. Dad opened the windows again as Lyle taxied back to the red barn he used as a hangar. I sat there, completely absorbed by the power and beauty of the flight.

That summer I sneaked away from our house whenever I could to watch Lyle working on his plane in the barn. If my parents wanted to find me, they always knew where to look. Lyle was patient, and always answered my questions. Sometimes he used a pencil to draw on a scrap of butcher paper, showing me the relationship between engine power and the airflow over the curved surface of the wings and the principle behind the ailerons, elevators, and rudder. At the age of four, I already had a basic understanding of lift and other aeronautical concepts.

Whenever Dad could get off work for a few hours, Lyle would take us flying. After the spring lamb season, midsummer was a slack period for a sheep farmer, but there was always work on a farm. So he'd tell his wife, Eva, "Shane wants to go flying" as an excuse to slip away from his afternoon chores.

After the first couple of flights, Lyle let Dad and me sit up in the front cockpit while he flew from the rear. Now I could really pretend that I was the pilot. With the engine howling on the takeoff roll and the pasture rushing past the windshield, I felt like the plane was under my control, even though I was tightly strapped to Dad's lap and under strict orders to keep my hands off the stick and throttle. But sometimes when we were airborne, Lyle would let me wrap my fingers around the front-cockpit stick while he banked left or right. Then I could actually feel the pilot's connection to the airplane.

After that summer, I had no doubt at all what I was going to do when I grew up. I would be a pilot.

And that stubborn determination did not change with the upheaval that hit our family later that year. Although their businesses were doing well, Mom and Dad had personal problems. My dad was a Vietnam veteran who had served in combat as an artillery forward observer, been wounded, and come home to South Dakota with two Bronze Stars, a Purple Heart, and a lot of traumatic memories. When my parents divorced, Mom took Lynnette and me back to her hometown of Norfolk in northeastern Nebraska.

My mother went to work as a nurse at the Veterans Administration home. She also tended bar at the country club and a place in town. She must have been exhausted most of the time, but never showed it. We needed the money, and she was determined that her children would not suffer. We certainly did not have any extras, but we never went hungry. Her example taught me as a very young child that I would have to work hard to achieve what I wanted.

Going to first grade in a "real" school after seeing Lynnette walk off each morning to the one-room schoolhouse in Loomis was a real thrill for me. I loved everything about Northern Hills, the chance to meet other kids, the teachers who read us stories. I could already count to a hundred, do some addition, and recite the alphabet because Lynnette had taught me at home in Loomis. But at school in Norfolk, I especially loved the brightly colored maps of the world that hung on the walls. Those were the countries I'd fly to when I was a pilot.

And I sure had not forgotten that dream. When Miss Anderson asked us what we wanted to be when we grew up, some kids said doctor or nurse or even major league baseball star. But my answer remained the same: "I'm going to be a pilot."

I was Miss Anderson's pet. She was a first-year teacher, and I was a skinny little kid from small-town South Dakota, with a struggling single mom, who always came to school early and loved to study. I was proud of my perfect attendance record until chicken pox kept me home for two weeks. Miss Anderson got the other kids to draw pretty get-well cards, which she sent to me with the individual homework assignments I worked on each day. When I got back to class, Miss Anderson hugged me and cried. She taught me what it meant to be dedicated to your profession.

That fall, our class ran a mock election. It was 1980. This was Republican farm country, and I ran the Ronald Reagan campaign. Reagan got it by a landslide. The local newspaper featured my picture on the front page. I was happy that our side won but understood we hadn't really worked very hard. And that was what I needed to do to become a pilot.

By second grade, I was taking advanced reading and math. Some kids were bored in school, and I thought that was a real waste because the teachers wanted to help us so much, the books were free, and when you looked at things, it was our job to study, to work just as hard as our parents did. I had fun on the playground and after school. But I always made sure my homework was done.

After second grade, Mom was able to move us out of the small apartment and into a little ranch house in the Woodland Park neighborhood. There were two or three basic models of homes on our street. Ours was definitely on the basic side. But I loved my new neighborhood. There were twenty-eight kids my age on our street alone. We all went to Grant Elementary School. I had made up my mind to achieve perfect attendance and wouldn't let a cold or a sore throat keep me home. I soon realized that math and science were my strongest subjects and took pride helping out the kids in my class who had trouble with them.

After school, we built forts in the shelterbelt of trees that ran along the creek behind the development. It was a safe neighborhood, and we could roam in packs, exploring the shallows of the creek in the summer for minnows, and tracking deer and raccoons in the winter snow, playing Indian scout.

But most of my summers were spent down at my uncles Bud and Mike's hog farms near Platte Center, south of Norfolk. I quickly learned the fine art of "walking beans," trudging down the endless rows of soybeans, chopping weeds with a corn knife. The work was pure drudgery. As the prairie sun grew higher in the morning, the bean fields became ovens. Biting stable flies rose in swarms from the hog barns, and there was no way to escape them.

After the first couple of days, I was miserable and wanted my mom to take me back home to Norfolk, where I could spend my vacation reading books in the shade or riding my chopper bike once the sun went down. But my two cousins, Uncle Bud's son, Junior, and Mike's boy, Rich, had no choice. They had no comfortable refuge to flee to. And although they didn't like the sun and flies any more than I did, they were farmers' kids from a tight-knit family. They worked. So did I. Those summers in the bean fields provided another valuable lesson: You learned something about your own strength and weakness when you stuck to a job you hated.

I also learned what was involved with hog farming. My first week at Uncle Bud's place, we had to castrate young boars, a squealing, blood-splattered ritual that taught me to respect where my food came from. Then that fall, I helped slaughter hogs for the first time, placing the galvanized tub underneath the hanging animal to collect blood for sausage. These were my people. This was where I came from. They knew about hard work, bad harvests, and low prices. But they also knew the importance of sticking together when times were tough.

By the time I was ten, I figured I had outgrown my little chopper bike. I wanted something with a motor on it. So I saved the pay I'd earned from my uncles and gift money from my mom and bought a light Yamaha GT-80 dirt bike. After school I'd roar along the dusty trail and down to the bed of the Elkhorn River and sneak into town to show the bike off to my buddies from school. It was illegal for kids my age to ride dirt bikes on the street, so if the cops spotted me, I'd take off for the riverbed again and roar home. But they knew where I lived, and were usually there waiting for me. They all knew how hard Mom worked, however, and always let me off with a warning.

One of my favorite pastimes was visiting with the vets at the VA home where Mom worked. Most of the men had fought in World War II, but there were still a few really old guys who had served in France during World War I. I loved to listen to their stories. I was old enough now to understand a little about what my dad had gone through in Vietnam, but World War II was still a mystery to me.

A red-haired guy named Hal would talk for hours about being a B-17 pilot in England. Many of the words like *Messerschmitt* and *Focke-Wulf* were completely unfamiliar to me, but I soon learned they were German fighters. Then Hal made me a beautiful model of a B-17 that he had painted with the exact squadron markings of his own bomber. I took it home and hung it from my bedroom ceiling among the several dozen other model planes and ships the guys at the VA home had made me. Some rainy afternoons, I'd put my toy soldiers in the bins of the scale model tractors and manure spreaders my uncles got from the farm equipment companies and push them into battle on my bedroom floor. Then I'd bring one of the warplanes down to strafe and bomb just like I'd seen in the old movies on TV. It always looked like a game, both on the TV screen and in my bedroom.

One old man named Bill built me a model of the battleship he had served on, the U.S.S. *Arizona.* I couldn't understand why it was so much bigger than the other model ships the vets had made for me.

"Where's this ship now, Mom?"

"It sank, Shane," she explained. "At a place called Pearl Harbor. It was bombed, and a lot of men were killed." Now I began to recognize a little about what being a veteran really meant. These kind, soft-spoken men had seen things more terrible than I could ever imagine. War had never been a game for these vets. They had served their country when it needed them. Now hardly anybody ever came to the VA home. They were left there day and night with their bingo and dominoes. I made up my mind then that I would visit as often as possible. It was the least I could do to honor their service.

About this time, my mom's boyfriend, Jerry Williams, moved in with us. He was friendly but strict. If Mom laid down rules, Jerry expected us to follow them. He became a second dad to me just when I needed one. This was definitely a good thing, because I was becoming something of a wise guy, and not everybody appreciated my sense of humor. Mom was still working as hard as ever. She'd come home after a full day at work, lie down for a fifteen-minute nap, and then head out for her night jobs. But I was expected to study hard and do my chores around the house. Jerry kept me on the straight and level.

By the time I finished grade school, I was getting tall but hadn't filled out much. I wore my hair in a way the girls seemed to like, short on the sides with a curly, sun-bleached mop on top. There were guys with thick necks and arms bigger around than my legs who told me I looked like some kind of a poodle dog. I didn't like to hear that and let them know. Luckily, my sister, Lynnette, was going with Tony Waugh, who stood six foot four and weighed in at around 270. A word or two from Tony was enough to convince these tough guys to lay off.

I decided it was probably a good idea to start working out myself. Since I wouldn't be intimidated by bullies, I had to be strong enough to discourage them on my own. I began lifting weights with the same methodical attention I tried to bring to every challenge I accepted. I'd also decided to work as hard as I could to earn my own money to relieve some of the financial pressure on Mom. She agreed, as long as my grades did not suffer. I went to work as a busboy at the Village Inn. Later, I took a job waxing and polishing combines at the Dinkel Corporation. It was outdoor work and paid well. Best of all, each big combine took hours to polish, a process that built up my arms better than any workout with weights.

The bodybuilding eventually paid off. I became strong enough to work as an orderly in the orthopedic ward of the local hospital during my high school years. This required lifting hefty patients and supporting people who'd had hip replacements as they walked the corridors. It was a satisfying job. But I never gave up my unofficial volunteer duty hanging out with the vets at the VA Center. Now that I knew so much more about the sacrifice they'd made for America, I considered it an honor just to be in their company. You never knew if some quiet newcomer sitting in the corner of the dayroom had won the Medal of Honor in the Normandy invasion or spent three years as a prisoner of war in the Pacific.

Since second grade, I worked my way steadily up from Cub to Life Scout. Now I had almost enough merit badges to become an Eagle Scout. But then one day at school a woman named Mrs. Askew gave a presentation on the Civil Air Patrol, an auxiliary branch of the U.S. Air Force, organized something along the lines of the Scouts. There was a big difference between the Scouts and the CAP, however. In the Civil Air Patrol, you learned how to fly, and could earn your private pilot's license by age sixteen.

When Mom got home from work that night, I told her how much I wanted to join. There was a forty-dollar fee, and you had to invest $140 in uniforms. This was money we just couldn't afford.

She looked at me a long time. "You really want this, don't you, Shane?"

"I sure do, Mom."

She sat down at the kitchen table and wrote a check.

The Civil Air Patrol absorbed me completely. For the first time since starting school, I sometimes felt distracted from my studies. But I soon realized the CAP made me want to study even harder than before. We had after-school classes in the history of flight, fundamental aeronautics, and space travel. And there were tests on what we learned in those classes. You advanced in rank depending on both your test score and your progress in military training. Discipline in one area, I now saw, was directly related to progress as a would-be aviator. And the reward for that progress was getting to fly in our squadron's Cessna 172.

The more skills I acquired, the more airtime I was given. One of the CAP's primary responsibilities is searching for downed aircraft, triangulating their position from the beeps of their Emergency Locator Transmitter radio beacon. So the other cadets and I dug in hard to the exacting tasks of map reading and applied geometry. Since math had always come quickly to me, I found this challenge relatively easy. And I also had the satisfaction of helping the girls and boys in my unit master this process.

My first year in the CAP, I tried to sharpen my military image, which was probably the weakest aspect of my squadron work. Mom taught me how to press sharp creases into my uniform, and when I visited Dad at his new home near Minneapolis, he showed me the finer points of putting a glossy spit shine onto my boots.

But there was still the problem of my hair. Although it was short enough on the sides to meet regulations, I had a narrow tail that I tucked under my collar. After drill, when the cadets went to hang out at Daylight Donuts, some of them kidded me I'd never make rank wearing that "rattail." I loved that haircut for the same illogical reason teenagers have always followed fads. But I realized keeping it might get in the way of my advancement in the Civil Air Patrol and possibly jeopardize earning my private pilot's license. I cut my hair. And I even became a drill instructor.

I also accepted the challenge of studying to become a tornado spotter. This, too, was important community work; we lived in one of the most active tornado belts of the Midwest. And when violent weather threatened in spring and summer, members of our squadron were often dispatched along country roads equipped with handheld radios to watch for and report funnel clouds snaking out of the black mass of the thunderheads.

Another specialty assignment I accepted was studying to become a warden of a nuclear bomb shelter. We had several in town, located in underground structures, which, given our proximity to the Strategic Air Command headquarters in Omaha, was a wise precaution. So, by age fifteen, I was a certified warden, and found myself guiding state Civil Defense officials around my particular shelter, clicking Geiger radiation counter in hand.

Because of the progress I was making, the squadron rewarded me with a 400-mile overland flight down to Omaha. I got to sit in the right seat and handle the controls under the supervision of the instructor pilot. *By next year I'll be flying on my own.* I knew by then that I wanted to become a military fighter pilot, and I recognized that the training the CAP provided could help me win an appointment to a service academy. And I did get to log about sixteen hours of instructed flight toward my license. But I'd have to go outside the CAP and pay for private lessons, a luxury I could not afford.

And, as things turned out, events intervened that took me out of the CAP. High school extracurricular activities pulled me in other directions. I played the electric bass guitar in the jazz band and the tuba in the big band. But my real love was football. The Norfolk High School

Panthers were a running team, so I was able to play both outside line-backer and wide receiver. Reluctantly, I left the squadron before I earned my pilot's license.

The May of my junior year at Norfolk High, the worst ordeal of my life began. Although I never want to repeat the experience, it taught me to value inner strength and not to quit when the situation seems hopeless.

My grandmother, Dorothy, had undergone hip surgery down in Lincoln, and my sister Lynnette, her fiancé, Tony Waugh, and I drove down in my mom's Nissan 200 SX one cool Thursday night after work to visit her the next day. It was a cramped little car, but the trip was only about two and a half hours, and when I took off the seat belt, I was able to tilt back the passenger seat and actually fall asleep. Lynnette was already curled up tight in the rear seat, sleeping.

But with the flat prairie highway and nobody to talk to, Tony got sleepy too. He dozed off as we were moving about sixty-five miles per hour with the cruise control set. The Nissan plowed straight into the back of a big John Deere tractor. The impact was so violent, we broke the tractor's rear axle.

The front of the Nissan was crushed up to the windshield. My head slammed through the glass, and my knees jammed into the firewall. Striking the windshield shredded my left eyelid and ripped my nose loose so that it flipped to one side, exposing bone and cartilage. The glass shards were like dozens of razor blades slicing my flesh and embedding in my scalp. Thankfully, I don't remember the actual moment of impact because I was asleep.

But when I regained consciousness, I was lying back in my seat and I heard a paramedic's distant voice speaking through the pulsing red strobe of the emergency lights.

"I think this one's dead," he said. He was talking about me. Somewhere unseen, Lynnette was howling with grief. But I couldn't answer because my lips were so torn. My stomach was steadily filling with blood gurgling down my throat and I groaned loudly before slipping once more into unconsciousness.

When I woke again, I experienced the most intense agony I've ever known. I was in the emergency room of the David City hospital. A doctor was sewing my left upper eyelid back together, while a nurse pinned down my arms. Because I had been unconscious and might have suffered a serious concussion, the trauma team could not give me general anesthesia. Instead, they relied on topical solution. But that wasn't enough to dampen the extreme pain of suturing my eyelids and face.

Finally, unable to control myself, I screamed with all my might. The doctor backed away. The nurse ran off and quickly returned with two orderlies, who restrained my arms with thick straps.

"Don't move," the doctor ordered. "You'll just make it worse."

Although I couldn't raise my arms, I gripped the bedrails so hard that I bent the stainless steel. Finally, the torture was over and I drifted off to sleep.

That evening my mom drove down to the hospital, passing the crash site on her way. The Nissan looked like a tin can that someone had stomped on. When she passed the door to my room, she did not recognize me. My face was a bloody mess, crossed with the black tracks of sutures, swollen more than double in size. But she saw Tony in the bed beside me. It was a miracle that neither he nor Lynnette had been seriously injured. As an experienced nurse, Mom did not want to tire me or strain my emotions, so she didn't stay in the room too long on that first visit. But, groggy as I was from the painkiller pills the doctors now gave me, I could see in her eyes how bad I looked.

When she left, I rang for a nurse. "Let me have a mirror," I said gruffly.

The nurse refused, but I insisted. Finally she brought a small mirror. I was sorry she did. My face was worse than I had feared. My left eye had been completely swallowed by blackened flesh. I looked like something out of a monster movie, a zombie or a werewolf who'd taken a beating with a baseball bat. When Mom entered the room and saw me with the mirror, she cried.

After the shock of that image, a cold, lonely fear began to form. My eyes were ruined. My face looked like a carnival freak show. *I'll never become a military pilot.* Hell, I would probably never be able to pass the eye exam for a private license. I was sixteen years old and my life was over.

After a day in the hospital, I was humiliated and angry. I hated having sponge baths and sipping soup through a straw between my battered lips.

"I'm leaving," I told my mom.

She saw I was serious. "You want to go home?"

"Absolutely not," I answered. It would be too painful to arouse the pity of my family and friends back in Norfolk.

Instead, Mom drove me down to Lynnette's apartment in Lincoln, where she was attending college. On the way we stopped at the wrecked car. With my one good eye, I stared at the jagged hole in the windshield and the smeared blood on the passenger's seat. It occurred to me that I would be dead if my knees hadn't jammed under the dash. But I felt little consolation.

The next week, Jerry Williams's son, Tony, came down to visit me. With the brutal honesty of teenagers, he said, "Man, Shane, you look really messed up."

"No kidding," I muttered. I had 220 stitches in my face, and a scalp riddled with glass shards. And every waking hour, the same thought returned. How bad was my left eye hurt? Will these injuries disqualify me from becoming a military pilot?

Naturally, I was also worried about my overall appearance; every teenager does. But the scars didn't bother me as much as the prospect of losing my dream to fly.

As the summer dragged on, my bruised knees healed. I began a series of reconstructive surgery procedures that slowly and miraculously hid the worst of the facial scars. But my left eyelid was still in bad shape as I started my senior year of high school. The suturing in the emergency room had saved the delicate flesh, but the upper eyelashes were tucked under the lid, sticking in, not out, so that as my eyelashes grew back, they constantly poked against my eyeball.

From preliminary tests, it appeared that my vision had not been compromised. But the ophthalmologists would know for sure only after the eyelid had been permanently reconstructed. I could either undergo the surgery at that time or wait to have it done after the football season.

I wanted to get back to something like a normal life. I chose to delay the surgery. Every few days for months, my mom helped me roll back the eyelid, and we used tweezers to pluck the backward-growing eyelashes that had become too irritating to bear an hour longer.

I'm afraid I took out a lot of my aggressions on the field that football season. I went from third string to starting as both wide receiver on the offense and outside linebacker on the defense. I hit people so hard that once during practice I even knocked myself out. I remember waking up with Coach Dan McLaughland smiling at me, saying, "Take it easy, Osborn. They're on our side!" But it gave me a way to focus my pent-up frustrations and anxiety about my face and eyes.

Finally, that winter after football season, I had the surgery. The doctor did an incredible job, slicing and tucking, and using practically microscopic sutures. When the wound healed, I had a normal eyelid.

The facial scars also began to fade as I regularly applied Retin-A to the worst of the puckered welts on my forehead. But I still looked pretty bad when I sat for my senior pictures, and they had to be sent to a specialist to be airbrushed.

That was small change. The crucial eye exam after the surgery showed that my vision had not been damaged.

"Are you certain?" I asked the ophthalmologist, Dr. Feidler.

"Your vision is perfect," Dr. Feidler assured me with a smile.

When Mom got home from work that night, the first thing she saw was my wide, confident grin.

That confidence was unshakable and had been a long time forming. God, I realized, had taken me through the terrible accident pretty much unscathed. He must have had a purpose for me in life.

"I'm going to fly," I said.

Between my sophomore and junior years, I had worked out hard every day, growing two and a half inches and putting on forty-five pounds, all of it muscle. By my senior year, I stood five feet eleven and weighed over 190. This was a change that people noticed. Apparently some of them who weren't used to working hard figured I had taken a shortcut to bulk up.

One day my football coach, Dan McLaughland, a former Air Force linguist who had flown on surveillance aircraft in the Middle East and was encouraging me to follow a military career, took me aside. "Shane," he said, "I don't believe this for a minute, but some of the guys are saying you must have taken steroids to get so big."

I felt my temper flare up. All those hours on the Nautilus machine and pumping iron and those lazy bastards accused me of popping steroids. "Well, Coach," I said, overcoming my anger, "let me match urine samples with them. I can provide the one I gave the Air Force to complete the application for the Academy."

When the coach spread the word, the rumors stopped. But it took a while for my resentment to go away. I had been exposed to drugs my whole life. No kid growing up in America when I did could have avoided them. By this point, there wasn't a safe neighborhood in our country, from the inner cities to the smallest farm towns, where some form of illegal drug or another didn't tempt young people.

That situation still prevails. But I had never felt that temptation, not for one minute. To me, drugs were cheating on life. They were a cowardly way of avoiding reality. Everyone I had ever seen who had gotten into drugs had lost some important part of themselves. If they didn't have the mental strength and discipline to stop quickly, it always cost them badly. Their lives were often broken, careers lost, marriages shattered.

That was why I was not tempted. Taking drugs had just not been an option. I had heard a lot about peer pressure, but that had always seemed like a lame excuse. Nobody really had to take that lazy way out. You can still have plenty of fun without messing up your mind and body.

And for me, there was a further consideration. I was determined to win a position at one of the service academies and go on to become a military pilot. I couldn't do that if I had drugs in my background.

My last year in high school, while still recovering from the car crash, I took the long competitive exams for both the Naval and Air Force academies. I did very well, but the competition for the relative handful of appointments allotted to Nebraska was fierce. Other, better-qualified young men and women were selected. But out of the blue, I was offered an appointment to the U.S. Military Academy at West Point, even though I had never applied. I was flattered by the offer but did not see earning a commission as an Army officer as providing a very certain route to becoming a jet fighter pilot.

Again, my dream of being a military pilot seemed to be over. I'd have to work my way through college and later pay for my own flight instruction. To a young man of eighteen, the disappointment seemed crushing.

Then the Navy contacted me. Because of my high test scores, the results of my interviews, and my high school background, I had been awarded a four-year Naval Reserve Officers Training Corps (NROTC) scholarship to any university that offered the program. The scholarship paid for tuition, fees, and books, as well as a small salary of $150 a month. If I successfully completed the program, I would be commissioned an ensign in the United States Navy. And if I did well as an NROTC midshipman, I'd probably be able to select Naval Aviation as my branch of the service and earn my Wings of Gold. I chose the University of Nebraska at Lincoln.

When I shared the great news with my mom, we both cried a little. She had worked so hard for so many years for me to reach this day.

In the morning, she went into the VA home and submitted her resignation. For fourteen years, she had tolerated an arbitrary and domineering boss with whom she had endured a constant, emotionally wearing personality clash. Now, with my leaving, she was free of it.

I was about to enter the adult world.

Chapter Two

# LUZON STRAIT

## 1 April 2001, 0800 Hours

I had not been able sleep in the curtained upper berth. Finally, still groggy, I dropped down to the deck and pulled on my LPU. By the return leg of the mission that afternoon, I'd be tired enough to rack out and sleep deeply. Now I had to relieve Jeff in the flight station.

Passing the galley, I grabbed a bottle of water, took a long swig, and moved forward up the center aisle of the tube. I stopped at Johnny Comerford's position, number 13, to review some details of today's mission that we'd already been briefed on before takeoff in the intelligence facility at Kadena.

"How's the gear working?" I asked, leaning close to his head to be heard above the engine noise.

"The backend's looking good, Shane," Johnny reported, indicating that our reconnaissance equipment had been tested and was functioning well. He had our trackline plotted on a laptop computer. "We'll be on track in about twenty-four minutes," he shouted.

Johnny was an extremely professional officer who had graduated from the U.S. Naval Academy at Annapolis in 1997 and been trained as a Naval Flight Officer. He was one of the most squared-away SEVALs I've ever worked with, not to mention one of my best friends. He was tall and slim and wore glasses, but his preppy appearance could hide the fact that he shared the same "work hard, play hard" attitude that I adhered to and valued in my friends. But we took our job as professional naval officers and aviators very seriously.

Johnny was also one of the smartest people I'd ever met, quietly confident and rigidly self-disciplined. His job was demanding, and he always rose to that challenge. I was grateful to have him on my crew during my first Det as a mission commander.

Passing the Nav station, I saw that Regina was again carefully updating the two inertial navigation systems by downloading into them the latest GPS position, which also gave us our ground speed. She did this whenever the inertials drifted off by more than a mile. By keeping our position current, we would arrive at the start of the trackline at the exact place and time we'd briefed.

On today's track, we would fly an irregular box pattern over the South China Sea that kept us well within international airspace. To a nonaviator, that kind of navigation might have seemed tricky, but we had been doing it so long that it was second nature. It was all just a question of confirming our position by GPS satellite and inertial navigation systems as we went onto the track heading, and then making the correct turns onto new headings at the required times, reconfirming our position each time we did so. This was like driving along invisible highways in the sky. Airliners did this every day. But they were directed along airways by ground air traffic controllers. We flew at odd-numbered altitudes, however, between the commercial airways the civilian jet lines used. Today, our track altitude was 22,500 feet.

When I returned to the flight station, Pat was writing down the latest update to the weather forecast, which had just come in over HF radio. I swapped out for Jeff in the left seat and studied the report.

Although the forecast Pat handed me on his clipboard remained good, I scanned the color weather radar located to the left of the center instrument panel. The screen was empty, the radar returns matching what we could see ahead: clear, sunny sky. But the outside air temperature hovered around 8 degrees Celsius, and was dropping. We were all well aware that dangerous icing conditions could suddenly occur at this altitude if we strayed into a cold cloud deck. And we were approaching the Luzon Strait between the Philippines and Taiwan, an area notorious for sudden, vicious weather, especially this time of year. This was the mixing zone where the colder air of the north Pacific met the tropical humidity of the South China Sea. Storms could pop up like mushrooms.

I definitely did not want to unexpectedly encounter another band of icing like the one we had struggled through a week earlier. After another early takeoff from Kadena, we had been on the long, slow climb to the south for a scheduled ten-and-a-half-hour mission. Once more we had been "bagged out," maximum gross weight, 142,000 pounds, over a third of that fuel. Again the weather forecast had been for optimal flying conditions, clear sky, and no turbulence. And once again the color radar had returned no indication of clouds or precipitation as we climbed, the engines set to give us a steady airspeed of 190 knots.

Then, at 16,000 feet, we had abruptly entered a layer of icy clouds that had not produced a bad return on the radar. But we immediately knew we were in trouble. The outside air temperature was around 3 degrees Celsius, ideal icing conditions.

Without waiting for orders, Petty Officer Wendy Westbrook reached up from the flight engineer's seat to the overhead panel, her hands jumping deftly among the switches of the electrical propeller and engine anti-ice system, which heated the metal of the props, engine air scoops, and inlets to prevent ice buildup.

The first five hundred feet of the climb through these clouds were uneventful. Then my entire windscreen suddenly became spider-webbed with frost like a picture from an old-fashioned children's book.

A moment later the nut on the windshield wiper had grown an ice ball the size of a grapefruit.

Wendy's fingers moved quickly across the bleed air panel, opening valves and flipping switches. Just as the name implied, the bleed air system bled hot air produced by the engine turbines and rerouted it through a complex web of ducts to melt ice already formed on vulnerable parts of the aircraft: the leading edges of the wings and tail and sections of the fuselage. This hot air should melt the rapidly building ice, but the bleed air also reduced turbine compression and robbed the engines of power.

Looking out the icy left windscreen, I was shocked to see the propellers on the numbers one and two engines had turned to white disks. I had never seen icing become so severe so quickly. Although Wendy had maximum bleed air working, the wings were getting heavier by the second from the thickening ice.

"I've got the controls, Jefe," I said, switching off the autopilot and turning to Jeff Vignery, who sat in the right seat. "This is very severe icing," I told him and Wendy. "We need to get out of this pretty quick."

But we had already slowed so badly from the drag caused by the ice that I could not maintain a climb. Instead, I leveled off at around 16,600 feet to let the airspeed build up. Still, the ice was forming faster than the bleed air or engine anti-icing could cope with it.

"We need to put on more power," I told Jeff and Wendy.

I took the power levers in my right hand and shoved them gradually forward from 1010 to 1049 TIT, Military Power. Ever so slowly, the plane began to climb again. But with the aircraft steadily accumulating ice, we would need every bit of thrust the engines could safely produce to maintain the climb. And we would also need to shed some of this ice.

"Maybe we can get above it," I said, then added to Jeff, "Get us cleared to eighteen."

The Japanese air traffic controllers immediately approved our request to climb to 18,000 feet.

I nudged the power levers a fraction of an inch farther forward. We had to maintain an airspeed of at least 200 knots in this climb. But we were still icing and slowing due to the ever-increasing drag. The major problem with trying to climb this slowly was the resulting high Angle of Attack with simultaneous high power setting. This invariably overheated the engines and provoked one or more Fire Warning lights.

If we got a Fire Warning light, we had to immediately shut down that engine and descend. More than once, EP-3Es or P-3 Orions had had to shut down an engine under such circumstances, abort the mission, and undertake a dangerous fuel-heavy three-engine landing.

Every flight crew's nightmare, of course, was that they would be hit with two Fire Warning lights from engines on the same wing while struggling to climb through the icy "goo." If that happened, controllability would become a real problem as the plane was forced to descend back into the frigid clouds.

But that morning we'd been lucky. Three hundred feet after increasing the power setting, we broke out into brilliant sunshine at around 18,000 feet. I was practically blinded by the dazzling white surface of the wings and the glittering disks on the props. We had cleared the unexpected icing layer just in time. In the dry sunshine, the ice began to melt.

As we continued our climb to cruise altitude, we scanned warily ahead and I kept a close watch on the weather radar. That had been one mission on which I had not managed to sleep.

The incident had been a reminder to us all that the sky was a dangerous place and military aviation was a hazardous occupation. You had to stay on top of events, or you could quickly get in over your head with catastrophic results.

• • •

I took the control yoke and switched off the autopilot with my right index finger. "I've got the controls," I said into my headset mike, formally announcing that I was flying. I wanted to get the feel of the plane's trim and balance before we went back on autopilot and got on track. Unlike many large planes, the EP-3E is inherently unstable at altitude and takes constant fine adjustment with the controls to fly stably when off autopilot. This is because the Lockheed Electra airframe had been shortened during the conversion to the P-3 maritime patrol plane in the 1960s. So we relied on the reliable old PB-20N autopilot to keep the plane straight and level during the long patrol flights.

We were still fuel heavy, and looking down I saw that Pat and Jeff had used the autopilot to rotate the wheel of the elevator trim tab a few degrees forward to help keep the nose level with the horizon. The trim tabs were small movable panels on the elevators, ailerons, and rudder that we could adjust without continually using the control yoke or rudder pedals. Later, when we burned off even more fuel, we could ease some of the nose-down trim.

When I switched back to autopilot, the plane settled into a solid groove. "Nice and smooth, Meeso," I said.

"Good to have a machine to do all the work," Pat replied.

Wendy Westbrook handed me a form that she'd just completed. "The fuel numbers, sir."

Even though I knew her fuel calculations would exactly match the quantity gauges on the center console, and that the fuel transfers among the five tanks she and Senior Mellos had performed since takeoff would have perfectly balanced the remaining load, I double-checked her work, comparing the numbers with the gauges. In the Navy, we followed procedure.

"Looks good," I said, handing back the form.

Any flight engineer mentored by Senior Chief Nick Mellos was certain to do things right. This was how Naval Aviation worked. Our squadron trained constantly, on simulators, on spare P-3C Orions (which had the same flight station as the EP-3E), and during Dets. This was what discipline meant.

As we droned southwest through the morning, I thought back to the years of discipline that had gone into my own training.

They say that basketball is the unofficial state religion of Indiana. If that's true, football fills the same role in Nebraska. The University of Nebraska at Lincoln is a great school, with high academic standards and plenty of social life, much of it focused on the football team, the Cornhuskers.

That was why my mom's decision to invest in an old lady's three-story house quite near the Cornhuskers' stadium showed her innate business sense. I could room there free, and we could rent rooms to my friends to cover the mortgage payments and insurance premium.

It took a fair amount of labor replacing carpet, hanging drywall, and painting. But, thanks to my brother-in-law Tony's handiwork, we had a livable house by the start of the fall semester in 1992. With four guys and two girls renting rooms, Mom's cash flow on the house was never a problem. And we also made money renting out desirable parking spots in the driveway and yard on home-game weekends during the football season.

And anybody who's been to a Cornhusker football weekend can tell you the campus turns into one big party. I worked hard during the week and had a blast on the weekends.

I had chosen math as my major. It was a subject I had always loved because of the logical precision involved. And although math at the university level was certainly harder than it had been in high school, I found the classes demanding but not overly difficult.

Naval ROTC also posed no mystery after my years as a Civil Air Patrol cadet in Norfolk. The university NROTC battalion was composed of three companies of midshipmen. As at Annapolis, the mids, both men and women, were ranked from midshipman fourth class (freshmen) to first class (seniors). The mids attended a weekly lecture and assembled every Thursday for drill. We also had mandatory study hours twice a week to ensure everyone was paying attention to academics. But for me, slacking off was never an option: If my grade point average dipped below 3.0 in my first year, I'd lose my scholarship.

The goal of NROTC is to produce well-rounded Navy officers who can acquire both professional and leadership skills. From my experience, the program accomplishes this very well, even better than the Naval Academy at Annapolis in some ways.

Plebes, as fourth-class midshipmen at Annapolis are known, lead a cloistered life. They're jolted out of bed by Reveille every morning at six-thirty and follow a daily schedule that is much more regimented than at a normal college. Their mealtimes are fixed. And there are mandatory study halls every night. Although Naval Academy midshipmen are granted more and more personal freedom as they progress, they are largely separated from the civilian world, in order to immerse them in a military environment.

As NROTC mids at the University of Nebraska, we certainly did not receive such concentrated military training. And we had to master the challenges of managing our own daily schedules, which involved shopping for food and clothes, and balancing our own checkbooks. When it came to waking up for an early class, that was our own responsibility. And some mornings were definitely harder than others. I partied plenty in college, but I always tried to keep my priorities straight. School came first, going out with the guys second.

Early in NROTC we mastered the Navy's basic personnel structure and the complete inventory of ships. This involved not just memorizing a list of letters and numbers, but also passing tests on the function and capabilities of vessels such as those comprising an aircraft carrier battle group. Then there were all the missile systems in our allies' and our arsenals as well as those of potential adversaries. I was particularly interested in surface-to-air missiles, as I was confident I would do well enough in NROTC to be selected for the Aviation branch and go on to become a fighter pilot.

Leadership was one of the essential elements stressed in the NROTC program. We had dozens of leadership exercises during my four years at the university. I particularly enjoyed the practical skill-building approach. We'd meet in the gym and divide into small teams to tackle specific challenges. I remember one that involved building a plank bridge between two sets of bleachers. Although the drop from the bridge down to the wrestling mats on the floor was not that great, the plank seemed awfully narrow. We had to guide people from one end to the other by swinging them ropes suspended from the ceiling. But the rule was that the assistants on the floor had to leave one by one as each person crossed successfully. That would have left the last person with no one to swing the ropes. So we had to identify the midshipman with the best sense of balance and the least fear of heights to volunteer. The value of that lesson was that we all needed each other, and that everybody had both strong and weak points.

By the second year of NROTC, we began to take on some of the responsibility for helping the incoming class of new mids. In many ways we were like a fraternity, although we had a number of women in the battalion. One of our responsibilities was to make sure these freshmen balanced their study load so their grades didn't suffer. As we advanced, we became unofficial mentors to the younger people who might have been struggling with subjects we had taken earlier.

Even though my class had assumed much of the responsibility for running the battalion by my senior year, we always maintained an open and unintimidating relationship with the younger midshipmen. "We're all middies" was the informal slogan. "We'll all get through this together." We worked hard together during the week, and we played hard together on weekends. This bond was especially tight among the prospective pilots. We knew we'd soon face the Navy's toughest selection process and we supported each other.

When I was a sophomore, the midshipmen interested in aviation went on a brief orientation trip to the sprawling Naval Air Station in Pensacola, Florida, site of Aviation Preflight Indoctrination (API), the boot camp for Navy pilots. Although we had only a few days during spring break, we did get some hands-on experience. Obviously the Navy wanted us to stay interested in aviation, so they made this orientation exciting.

A smiling staff of experienced petty officers introduced the mids to a weird apparatus affectionately known as the "Spin and Puke." This was a big drumlike structure sitting mysteriously on the concrete floor of a nondescript one-story building. It might have been a giant's chocolate cake perched on a lunch counter lazy Susan with thick electrical cables snaking out from underneath. But there were six narrower vertical cylinders, each big enough to accommodate a single person, spaced evenly around the drum's inner wall. The narrow cylinders rotated with the bigger drum, but they could also spin in the opposite direction, then reverse their spin when the drum changed direction. When the doors were closed on both the smaller cylinders and the large drum, you lost all reference to the outside world.

That was the idea.

"Next group of six," an instructor called.

I marched in with five other midshipmen.

"Take a tube," the instructor said. "Doesn't matter which one."

I entered mine, wary there might be some residual stench if someone had in fact barfed earlier. But all I could smell was the faint ozone of electrical controls piping through the overhead air-conditioning vent.

The instructor closed the sliding door. I felt the large cylinder begin to move slowly to the right. Then my tube also began to turn. But was I moving right or left? The tube slowed and changed direction. Was I still moving right? I thought so, but I couldn't be certain. We had been told to keep our eyes open to prevent nausea. I did as I was told. But the cylinder and the drum kept weaving through this sickening tango. One minute I was sure I was spinning to the right, the next I knew I had stopped completely. But if I blinked, it felt like I'd been moving steadily left all along.

I got through my spin fine, but some of the people felt a little woozy.

After all the mids had been through the Spin and Puke, a lieutenant gave a talk on how our sense of spatial orientation depends on nerves, fluid-filled canals, and a collection of tiny, delicately balanced bones in our inner ears. When we cannot match the sense of motion with visual reference points, we lose our equilibrium. This extremely hazardous condition can occur when pilots fly into clouds or the horizon becomes obscured and they have no instruments such as the gyro's artificial horizon or the Vertical Speed Indicator to orientate them.

The problem is called vertigo. It can kill you fast. You think you're flying straight and level when you may be banking nose down or pitching up toward a stall. I knew this from the CAP. But the session in the Spin and Puke reminded me.

"This is a little lesson for you," the lieutenant said. "Trust your instruments, not your senses. Instruments never lie, but as you've just found out, your senses can fool you."

Now we got rides aboard the Navy's primary flight trainer, the two-place T-34C Turbomentor. With straight wings, rakish nose, and a bulging Plexiglas canopy, the single-engine trainer resembles a World War II fighter. The plane is highly maneuverable and has dual flight controls and duplicate instruments in each of the tandem cockpits.

The morning I took my ride in the bright orange-and-white T-34C, I was excited, hoping the instructor pilot, a lieutenant from nearby Whiting Field, might actually do some aerobatics once we were at a safe altitude.

But as we walked through the preflight inspection, I wasn't sure he would. Although friendly enough, the lieutenant looked like a straight arrow who no doubt had more interesting things to do with the day than to take some overeager midshipman flying along the Gulf beaches. I couldn't have been more wrong. This was a guy who flat-out loved to fly. And he also loved sharing that joy with the kid in the backseat.

After a short, fast takeoff and a smooth climb-out, the instructor headed for a block of training airspace, where he could put the plane through its paces. It was a sparkling Florida April morning. A lot of my fellow college sophomores on the beach during spring break were nursing hangovers or engrossed by the local bikini talent. Naturally, I would have liked to have been there myself. But just then I was exactly where I wanted to be. There'd be plenty of time to drink a few cold ones on the beach after this flight. Sitting behind the instructor pilot, I relived those flights in Lyle Brewer's little Piper Cub all those years before. But the T-34C was a lot more airplane. It was fully aerobatic, which meant it could fly the same maneuvers as the Navy's hottest fighters, just a lot slower.

"How 'bout we try an aileron roll?" the instructor suggested.

Before I could reply, he slid the control stick precisely 90 degrees right and we were standing on the right wing. A moment later, the stick rotated another 90 degrees. We were inverted, and I was hanging in my harness, looking up at the ocean. The instructor continued the roll in two more 90-degree slices until we were back to normal flight, wings level.

"Feeling okay?"

"I feel great, sir," I said into my helmet mike.

"Well, then," he said, "now we'll do a couple loops."

He put on some power. The turbine whined and the prop whistled as the lieutenant hauled back smoothly on the stick. The nose rose steeply above the horizon, and I felt suddenly heavy in my seat. *G forces.* This was *so* cool. We were over the top, briefly inverted again. We almost seemed to pause in midair, then we were diving steeply. This was a lot more fun than droning across Nebraska in a Cessna 172.

But the morning wasn't over. We did a few more rolls and loops, and the lieutenant innocently asked, "They still show the mids those World War II films?"

"Yes, sir."

"Well, hey," he sang out happily, "let's go strafe an enemy beachhead, then."

Before I could answer, he snap-rolled right and was diving toward the broad white band of the tourist beach to the east. The land and the sea swept toward us in a green-and-blue blur.

"Enemy landing craft, one o'clock." He turned the nose slightly seaward and continued diving. "Bam! Bam! Bam! Dah! Dah! Dah!" he shouted cheerfully like a kid mimicking the machine guns of a Corsair in the Black Sheep Squadron. "We got 'em."

*Are all instructor pilots like this?* I wondered. If they were, I wanted to go to flight school even more.

Little did I know.

• • •

As I entered my fourth year at the University of Nebraska, I realized I needed the equivalent of almost four semesters of a heavy credit load to complete my math major, specializing in statistics and actuarial science. In fact, it is common for people in this field to require five years to complete the degree. But two of my best friends from the NROTC battalion, Lance Wiese and Lonny Beberniss, had been commissioned as ensigns, selected for Aviation, and were already in flight training. I did not want to wait any longer than necessary. So I almost doubled up my credits and concentrated on the most difficult math courses. Luckily, my math professor, Dr. Steve Dunbar, was sympathetic and spent a lot of extra time with me. By spring 1996, I knew I'd have enough credits to graduate in May. And my grade point average—including the all-important NROTC grades—had remained high.

Now I had to sweat out selection. I hadn't joined the program to become what Naval Aviators call a "boat driver." Everybody who'd ever met me knew how badly I wanted to fly. In fact, the battalion skipper, Captain Stephen Delaplane, later told people, "Shane needs to fly like most people need to breathe."

So, I was really anxious about approval of my selection choice when the skipper called me to his office. He had a grim look on his face as he handed over a letter on Department of the Navy stationery.

"Here are your orders, Shane," he said, shaking his head. "I'm really sorry." Captain Delaplane was a tough old sailor, a diver who specialized in recovering explosive ordnance in murky harbors. The news must be really bad to make him look so gloomy.

My eye jumped across the orders, and I felt a sagging disappointment hot beneath my throat. I was to report to a Navy office in the Pentagon Annex for duty as a deputy to an assistant to a "coordinating chairperson" of an alphabet soup personnel committee engaged in some form of obscure bureaucratic endeavor that I'd never heard of. In other words, I'd be riding a desk in a basement for the next four years while my buddies qualified as Navy fighter pilots and joined the fleet on carriers. I was stunned, and for once didn't have anything to say.

But when I looked up, I caught a twinkle in the skipper's eye. He didn't have the heart to stretch out this prank any further. He came around the desk and gave me my real orders. "You got your pilot slot," he said. "You're going to Pensacola."

My Aviation Preflight Indoctrination would not begin until October, however, so I was able to pass a great summer. The highlight was a thirty-day cruise aboard the aircraft carrier U.S.S. *Kitty Hawk* while still a first class midshipman. Life aboard an 80,000-ton fleet carrier conducting flight operations day and night in the Pacific was exciting. More than 5,000 sailors lived aboard this self-contained floating city. On certain decks, you could stand in a passageway and stare down the entire length of the ship, 1,000 feet from bow to stern.

Aboard *Kitty Hawk*, I experienced the incredible thrill of a catapult launch, first in a backseat of an EA-6 Prowler electronic countermeasure jet, then in an S-3 Viking patrol bomber. Although I'd seen cat launches countless times on TV and in movies, it was an altogether different experience to be strapped tightly in your seat, staring out at the gray-black flight deck as the engines screamed, revving up to full Military Power just before the steam-powered cat piston fired, flinging your plane into the air. You went from a standstill to a howling climb in about two seconds. *That* was pure exhilaration.

Whenever I had free time during Flight Quarters, as the Navy called flying operations on a carrier, I always found a place to watch the fighters launch and recover. I had probably seen the movie *Top Gun* half a dozen times, but standing on the observation platform as the huge ship plunged into the chill north Pacific wind at over 30 knots made carrier flying much more vivid than any film. Brine from the bow wave rose in clouds. The engine exhaust from the F-14 Tomcats and F/A-18 Hornets wafted up, giving the wide flight deck an industrial, bus-station stench. Crewmen in different colored jerseys that identified their specialties scurried about, some within inches of the big fighters' engine sucking inlets or hot exhaust pipes. When the planes were positioned on the catapult for launch, the aircraft came alive and seemed to hunker down, crouched beneath the unleashed thrust of powerful fanjets driving at afterburner.

Then the cat fired and those sleek gray fighters screamed away, driven by pulsing tongues of afterburner flame. The noise and the raw energy of the launches always overwhelmed me.

But, if anything, cable-arrested recovery on the flight deck was even more spectacular. An approaching Tomcat would appear in the distance, hanging in the sky like a tiny gray bird. But that was only an illusion. Within seconds, the plane would double in size, then double again. And you could see how fast the 30-ton fighter was really flying. Suddenly the crooked steel finger of the tailhook hanging from the fuselage became visible. The Tomcat would drop toward the deck as if the pilot were intentionally crashing into the ship. And that's basically what a carrier landing is, a controlled crash on the flight deck in which the tailhook snags one of the four thick steel arresting wires and is slowed to a stop with the same speed and violence of a catapult launch.

At the moment an aircraft slammed down, everyone watching would catch their breath, waiting to be sure the pilot snagged a wire. Pilots who hit the deck too far forward of the arresting gear were called "bolters." They had about one second, max, to throw their throttles to afterburner and scream down the remaining length of the flight deck, clawing for takeoff speed without catapult assistance. If they hesitated, the plane would stall and crash into the sea. Lucky pilots would be able to eject and their chutes would open to carry them free of the ship before they were run over.

This was the most exacting, unforgiving type of aviation in the world. Carrier-qualified Naval Aviators truly were a special breed. I was determined to become one of them.

At the end of the cruise, I joined a group of sailors aboard a C-2 Greyhound cargo plane for a flight from the ship to Pearl Harbor. Now I finally had the chance to see the place where the actual U.S.S. *Arizona* rested. And when I visited the ship resting in the oily water of the old Battleship Row on Ford Island, I though of Bill and the other vets back at the VA home in Norfolk. Back in Nebraska, Rear Admiral Pierce Johnson, my mother's cousin, commissioned me an ensign in the United States Navy on my birthday, June 21, 1996. I was an officer and felt confident I would earn my Wings of Gold and be a Navy pilot within two years.

When I reported to Pensacola for API, I joined a group of thirty student officers assembling to begin the long, difficult road to becoming Naval Aviators. But we had a few weeks before classes actually started. So my apartment mates, Ensigns Al Madrid, Mike Cooper, Jenny Mraw, Heather Barackman, and I began a serious reconnaissance of the beach bars along the strip. API was going to be tough. Our byword became "The time to party is now."

Rumor had it that there was a "rollback" rate of over 50 percent in API. This meant at least half our class would not finish the six-week program on schedule and would be held back for additional training.

Eventually, some would wash out and transfer to another branch of the Navy, but our instructors made it clear that choice would be ours. They would give us every opportunity to finish API and go on to Primary Flight Training.

But first we had to pass the toughest challenge to becoming a pilot, the incredibly rigorous Navy flight physical, affectionately known as the "NAMI WHAMI." This ordeal lasted two long days, during which vampire corpsmen practically bled us dry taking samples for lab tests. Unsmiling orthopedists used chilly steel tapes to measure our spines and legs, searching for any skeletal imperfection. They made us duck-walk, and then checked our thigh muscles for spasm. Next we had to hop on one foot and repeat the exercise with first one, then both eyes closed.

I was nervous about the eye tests, even though I had repeatedly been tested with perfect vision after the car accident and reconstructive surgery. But a candidate pilot's eyes were the weakest point. There was even a legend among Aviation cadets that you could blow the exam if you strained your eyes by reading practically *anything* up to a month before the NAMI. So, I had avoided reading whenever possible, and even cut way back on watching television.

The second morning, after a couple of hours swallowing chalky liquid barium and being X-rayed on tilting tables, we were delivered to the eye and vision specialists. They had equipment you'll never encounter at a normal hospital. One was a machine that used a computer to carefully measure the curvature of each eye, both external and internal. The device searched for any imperfection or evidence of injury that might not withstand the strain of high-G flight. Beyond the standard chart-reading tests, we were also subjected to intense depth-perception exercises. Staring into an enclosed apparatus at a field of meaningless colored geometric shapes, we had to signal the exact moment one of them seemed to move farther or closer to us. Once more, I sweated out the results of an eye exam. But this one had been far more demanding than anything I'd undergone before.

Word came within two days: I had made it through the ordeal of the medical exams, but several guys weren't so fortunate. They left Pensacola to be trained for other service in the Fleet.

I did not find the individual ground school classes in API particularly difficult. And subjects such as Aerodynamics, Propulsion, and Weather really intrigued me. This stuff was no longer theory, but information we would soon put to practical use in Primary Flight Training.

The instructors bombarded us with the material at a furious pace. And we were expected to absorb this increasingly complex information the first time around. I tried to maintain the study discipline that got me through my marathon final academic year at Nebraska, so the API classes presented no real challenge as long as I worked hard.

But the physical and water survival training proved surprisingly tough. I had continued to lift weights all summer, so I was in excellent shape. But I've never been a great distance runner. And one of our first PT challenges was a one-and-a-half-mile run through sand and brush, which we had to complete in 10½ minutes. We had a pair of these runs the first week. I made it across the line just on time in both runs.

Our survival swim training classes were held in the huge echoing indoor pool. The instructors evaluated our basic swimming skills very carefully. The few who did not seem very comfortable in the water were taken aside for remedial training. The rest of us moved on to the syllabus. We learned the art of survival floating, lying on your back and fluttering your extended hands just enough to keep your face above the surface.

That was no problem in swim trunks. But now we had to conduct our water survival training the way we would fly, wearing a flight suit, helmet, steel-toed boots, Nomex gloves, and SV-2 with the LPU uninflated. The survival float and treading water became a real endurance test. The killer was the final hurdle: a five minute survival float followed by treading water for five minutes until your thighs were burning from kicking the ten-pound boots. When the instructor blew his whistle, we had to somehow find the breath to orally inflate our LPUs.

That exercise really took it out of me. I knew it would because I'd had to take a similar test twice preparing for the summer cruise aboard the *Kitty Hawk* in order to get a ride off the boat. After that, I spent a lot of time in apartment house swimming pools, dressed in a flight suit, as I practiced the survival float and treading water prior to being graded on the exercise for API.

The classes intensified. The Navigation instructor expected us to practically memorize the text during the one-week course. Tests were now eighty questions with eighty minutes to complete them. I kept my average scores in the nineties.

We went through the high altitude chamber, a claustrophobic, welded-steel tube that looked like a railroad tank car. Seated on benches, the cadets wore oxygen masks, while the instructors took us "up" to an altitude of 35,000 feet by depressurizing the chamber. Breathing pure oxygen in this harsh environment, I felt normal, perfectly aware of my surroundings. Then the instructors took us back "down" to 25,000 feet.

"Masks off," the loudspeaker ordered.

We followed the order. Now came the manual dexterity tests we'd been briefed on. The Pensacola Patty Cake was the most demanding as your brain became starved of oxygen. Like three-year-olds in preschool, we tried to pat the open palm of the guy or woman next to us, alternating left and right hands. The first twenty seconds or so this was no problem. Then my partner, a good friend named Cliff Allen, hit my left hand twice. I stupidly followed his move, suddenly thinking this was a good idea and that the rules had changed. Ninety seconds into the exercise, we both had to stop and start again.

"Numbers ten and eleven," the instructor ordered, "masks back on." We had become hypoxic, thinking we were doing fine when we looked like a couple of drunks, which was no doubt nothing new for Cliff and me. That was another vital lesson: Don't underestimate the danger of high-altitude flight.

Our final week of water survival training involved the helo and the Dilbert dunkers. The helo dunker looked like a big gray soft-drink can lying sideways on a ramp above the lip of the pool. The apparatus simulates the cabin of a helicopter crashing at sea. Eight of us at a time strapped into seats on either side, and the dunker lurched down the rails into the pool. Water rushed in the open windows, the cabin rolled over at a depth of about eight feet. You had to keep from panicking while you unbuckled your harness and worked your way along the fixtures on the cabin walls that we'd been taught would lead to the assigned exit hatch. I didn't have any problem with this, although some of the guys were claustrophobic and freaked out, thrashing wildly and using up their breath. There were scuba divers just outside the helo dunker waiting to grab those people who seemed disoriented.

But the simulated night exercise in the helo dunker was much tougher. And scarier. We wore blacked-out goggles that blinded us and we had to switch from the seats we'd originally used and grope for a different escape hatch from the one we'd previously used.

On my first night dunking, I lost orientation and found myself probing along the curved metal wall in the wrong direction, kicked by the unseen struggling cadets around me. *I'm not going to make it,* I thought, certain I couldn't hold my breath long enough to find the escape hatch. I had to fight the urge to rip off the black goggles and swim to safety. But I swallowed the panic, reversed course, and pulled myself to the rectangular hatch. When my head broke the surface of the pool, I gasped, gratefully filling my lungs.

We had to successfully exit the helo dunker four times in six trials or we would keep repeating the exercise. Those who couldn't master the technique got rolled back. But I had passed the test.

In contrast, the single-place Dilbert dunker was much less challenging despite the hype it had received in the movie *An Officer and a Gentleman.* Meant to simulate a fighter cockpit, the dunker slammed down rails at a steep angle and flipped upside down once underwater. I had no trouble getting out, but I did end up with a noseful of harshly chlorinated water.

Our water survival training ended up with parachute exercises out on the bay. We were towed in a parasail behind a powerful speedboat up to 500 feet, and then dropped the towline and floated down to the water. Once we hit, it was off with the chute harness, clear yourself of the canopy before it dragged you down with it, and wait for our helicopter rescue. The H-3 clattered to a deafening hover above me and dropped an inflated yellow rescue collar on a winch cable. Then the winch operator lifted me from the bay like a big green, dripping fish. The key to getting lifted all the way up was to bribe the person running the winch with a candy bar or some other snack. But in the best traditions of API, the helo crew dunked me back in the saltwater.

After a welcome break over Christmas and New Year's, during which I stuffed myself on my mom's and grandma Dorothy's rich Nebraska cooking, it was back to Pensacola to complete API. I now was confident I wouldn't be rolled back. In fact, I managed to complete the course with one of the better cumulative scores. But eighteen of the cadets who had started API with me in October had been rolled back.

Now I had to decide on where I'd take my Primary Flight Training in the T-34C Turbomentor. The course was offered at Naval Air Station

Whiting Field adjacent to Pensacola and at Naval Air Station Corpus Christi, Texas. After the drudgery of API, I wanted a change of scene. So I chose Corpus Christi.

First, the surviving members of my class headed for New Orleans and spent Mardi Gras in the French Quarter. After the focused intensity of API, we really cut loose. This was the biggest party in the country, and it was great to be out there boogying until dawn. Then it was off to Corpus for Primary.

There are two training squadrons at NAS, Corpus Christi, VT-27 and VT-28. Although they are almost identical in number of instructors and they both have the same access to aircraft and flying hours, I discovered there was a world of difference between the two.

I was either fortunate, or unlucky enough, to have drawn VT-27, depending on your point of view. VT-27 was reputed to be the toughest training squadron in the Navy and was unofficially called the "Flying Luftwaffe" for the instructors' Prussian discipline. The atmosphere at our sister squadron, the VT-28, "The Flying Club," was a lot more relaxed.

In Primary, you lived or died on your scored performance. From the first four weeks of ground training through soloing and the months of specialized flying curriculum in the T-34C, we would be graded on everything we did in class and in the airplane. Grades ranged from Above Average ("Aboves"), to Average, Below Average, to Unsatisfactory ("Downs").

We slogged through T-34C Aircraft Systems and Familiarization just like countless cadets had before us. Sitting in simulators called Cockpit Procedure Trainers (CPTs) we listened intently while a bored civilian instructor introduced us to the flight controls and instruments and reviewed the plane's systems. Hour after hour, we went through engine start-up and shutdown checklists. We raised and lowered the flaps and landing gear. We calibrated the altimeter for different barometric pressures and changed radio channels. After CPT-1, the first week of these sessions, we were required to string together the procedures we had learned so far into proper sequence. By the skin of my teeth I got my first Above Average in VT-27.

This was a small victory. The CPT-2 and -3 syllabi involved correctly reacting to emergencies that the instructors began to throw at us. These ranged from engine fires to wing- and propeller-icing to sudden drops in fuel flow or loss of hydraulic pressure. I scored only Average in several of these procedures tests and vowed to do better.

I was determined to accumulate as many Aboves as I possibly could. I would need them to make "jet grades" on the Naval Standard Scores required for Advanced Training in fighters. Only a small percentage of Primary cadets earn these grades and eventually become qualified in the F-14 Tomcat, F/A-18 Hornet, or other carrier-based aircraft. The rest are selected for lower performance aircraft, with many shunted off to helicopters. I recognized and accepted the competition I faced and was confident I could meet the challenge.

The principle in Primary ground school is developing cumulative knowledge. The instructors introduced us to basic information the first week, then expected us to put that knowledge into practice, each week building on what had come before. Mastering emergency procedures—which came in the simulator just as fast as they did in a real cockpit—depended on that foundation. The final test had me sweating.

But I passed with a high score, raising my number of Aboves. However, the only Aboves that counted toward your cumulative score were those you earned in the airplane or in simulators during the next phase of training.

By the time I completed the four weeks of the familiarization course, I was beginning to understand a little about the approach the Navy took to forming aviators. Now I was ready to go flying.

In contrast to the joyride I'd been given as a visiting midshipman, flying in an actual training squadron was an infinitely more structured situation. The Familiarization Flights (FAMs) leading to solo all involved the student's personal instructor, known as the "on-wing." Mine was Lieutenant Jeff Nelson, one of the most skilled but also most demanding instructors in VT-27. He was known to be abrasive with lazy cadets and had absolutely no tolerance for those who didn't deliver top performances.

Each FAM was structured around a detailed syllabus that included a list of items on which we would be scored. If we accumulated two Unsatisfactories or "Downs" during our FAMs, we were invited to visit the training wing commander for a severe dressing-down and possibly attrition from flight school. That would reflect badly on the squadron even though the Luftwaffe didn't seem to care. They expected us to study the syllabus hard before each flight, and be prepared to put theory into practice.

After the heavy thunderstorms and coastal flooding that had plagued our month in ground school, I showed up on the flightline for FAM 1, my first flight with Lieutenant Nelson. In our first meeting the day before, he had referred to a clipboard that listed my API and CPT scores, as well as my sixteen hours of instructed flight as a high school kid in the Civil Air Patrol. Then he walked me out to a hardstand and through a quick preflight on the T-34C we were scheduled to fly the next day. He immediately began throwing questions at me, pointing at the curved exhaust pipe and demanding to know the gas temperature or checking to see if I knew the proper inflation pressure of the tires. Luckily I had answers that seemed to satisfy him.

The next morning he had me conduct the preflight walk-around and explain every item I was checking. Once more I seemed to pass muster. I climbed into the front cockpit, carefully buckled my seat harness, making sure my parachute ripcord was not snagged, and closed my canopy. Lieutenant Nelson immediately called for the before-start checklist, and I began to snap off the items, trying to keep up with his staccato responses. With the Texas spring sun beating down, it was stifling under the Plexiglas canopy, and there was so much sweat in my eyes, I could hardly read the checklist.

Once we had the engine running smoothly, the air conditioner kicked in and cooled the cockpit. Lieutenant Nelson called for taxi clearance and we joined the line of other Turbomentors working their way toward the end of the long runway. The sky above the field looked awfully crowded. But I knew there was a lot more room in the training area where we headed. Lieutenant Nelson ran up the engine for temperature, shaft horsepower, and RPM checks, and then got our takeoff clearance from the tower.

"Watch how it's done," he said into his helmet mike. "You'll be taking off tomorrow."

As on my first T-34C flight, the takeoff run was short and the climbout smooth. We reached the training area above Padre Island at an altitude of precisely 5,000 feet.

"Give me a shallow three-sixty, five-degree bank angle, to the right," he ordered.

I nervously eased the stick right, while neither pushing forward nor pulling back so that the nose did not rise or fall. But I didn't succeed. Lieutenant Nelson's hand was on his stick abruptly, correcting my mistake: I'd let the right wing drop below five degrees and the nose dip. We had lost fifty feet of altitude.

"If you want to fly in the Navy, Ensign," he said harshly, "you've got to do better than that."

My mouth was dry, my face burning. *He's going to give me a Down.* But Lieutenant Nelson was just probing to see if he could rattle me. Somehow I completed the turn and got back to our original heading at the proper altitude.

That night I studied the FAM 2 syllabus for hours. I had to memorize every detail of ground procedures and takeoff. And the effort worked. The next morning I found myself sitting in the aircraft with the engine throbbing as I checked my instruments a final time and advanced the throttle.

The plane howled down the runway centerline as my eye jumped from the engine and airspeed instruments to the horizon. The rudder pedals came alive beneath my feet, and the plane just seemed to lift off by itself as we passed 100 knots. For the next few minutes, I was actually flying, retracting flaps, and talking to the tower for my departure heading all at the same time. Then Lieutenant Nelson took the controls and we went back out to the training area to practice level banks and turns and climbs and descents between specific altitudes.

After the landing, which he flew, Nelson told me to taxi back to the hardstand and shut down. Walking back to debrief the flight, he had a clouded look on his face.

"Ensign, you didn't learn how to taxi an airplane like that in just sixteen hours with the Civil Air Patrol." His voice was grim. "I'm going to find out just how much you've actually flown before."

"That's all the hours I have, sir." Lying about this matter would be a breach of the Honor Code, a way to unfairly get a leg up on my fellow cadets.

Lieutenant Nelson must have contacted a CAP officer back in Norfolk who verified my total flying time as sixteen hours of instructed flight in a Cessna. After that his attitude toward me softened somewhat. But he was still an exacting instructor.

Each FAM was exhausting. As in ground school, I had to bring together everything I had learned before. On FAM 3, I landed the aircraft. It probably wasn't the most graceful landing a cadet has ever flown, but I knew it wasn't the most awkward. Lieutenant Nelson noted that I had been about five knots too fast on touchdown. But he still gave me an Above for that part of the flight. The next day, however, I got my first Below Average score when I completely missed the landmark of the small rock jetty at Fish Pass during route-checking procedure.

"Look, Ensign," Nelson said, "if you really want to do this, you need to get into the books and study what we brief."

Indeed, he had given me a small hand-drawn map of the training area the day before with the landmarks we'd be using in the morning. The mistake had been mine. You didn't get too many second chances with Lieutenant Jeff Nelson.

Now a frustrating pattern developed: I would study the syllabus intensely, only to have the next day's scheduled FAM scrubbed due to bad weather or lack of airplanes. There were a lot of students further ahead in the curriculum than my group, and they took priority. So we just had to wait our turn. This meant there could be as long as ten-day delays between FAMs. This was how the FAMs went that summer. Each was a grueling physical and mental challenge. How could I coordinate the hundreds of procedural details with my on-wing's inflexible requirement to fly with crisp precision when my flying skills had gotten rusty due to the waits between flights? But somehow my skills did accumulate. I suppose this involved the equivalent of the visualization techniques that high-performance athletes such as Olympic skiers employ. Like them, I studied every aspect of the next day's problem intensely until I could actually picture myself performing the tasks. This reduced the challenge to manageable proportions.

Within a month, I was landing and taking off routinely and flying the crowded flight pattern around the field with something like a relaxed attitude. But when I got too relaxed, the syllabus and Lieutenant Nelson would haul me short.

We began Basic Aerobatic Maneuvers, not just the benign loops and rolls I'd earlier experienced, but recovering from spins. A spin is every young pilot's nightmare. In a spin, the airplane usually stalls and departs from controlled flight and basically becomes a falling object. From inside the cockpit, you look straight down and watch the earth twisting toward you in a sickening blur. If you give in, you're dead. The T-34C does not have ejection seats, and centrifugal force makes it almost impossible to bail out during a bad spin.

But, as with everything else in Primary, there was a procedure to deal with the emergency. Correctly coordinating ailerons, rudder, and elevators dampened the spin. But the maneuver had to be immediately followed by a 3-G pullout. After my on-wing demonstrated this a couple times, he leveled off at 8,000 feet and said, "You try it now."

"Roger, sir," I replied uncertainly, taking the controls.

Once more my technique was not as pretty as it could be, and I was a bit harsh on the pullout—my jaw sagged for a second or two from the G load—but Lieutenant Nelson conceded it was a "good" effort.

Next, we concentrated on Basic Instruments, the vital cockpit displays that told us the aircraft's "attitude," its position in the sky relative to the earth. The gyro showed us if the wings were level and whether the nose was above or below the horizon, while the Vertical Speed Indicator revealed how fast we were climbing or descending. It was one thing to practice with these instruments in a ground simulator, but something else altogether to be sitting in the hooded rear cockpit deprived of outside reference points, banking, climbing, and turning while Lieutenant Nelson barked orders.

The weeks passed with the inevitable weather delays and unavailability of airplanes. By FAM 13, I had successfully learned how to deal with in-flight emergencies, ranging from an engine fire to jammed landing gear, to loss of power at high altitude. On the debrief after that flight, Lieutenant Nelson smiled. "You're going to solo tomorrow."

On the past several FAMs, I had been flying from the front seat, and my on-wing had kept his comments to a minimum. But I'd always been aware he was there behind me to take over in a real emergency. Now I was going to be on my own.

On August 23, 1997, I methodically worked through my preflight checklists and taxied to the end of the runway, listening to the rapid chatter of radio traffic in my earphones. I was thirsty, but had not wanted to drink too much water because I had to log a full hour and a half to make this solo flight count. Fighting the illogical urge to look back for the stern, comforting presence of Lieutenant Nelson in the rear cockpit, I went to takeoff power and was wheels-up less than halfway down the runway. It was a hazy, humid Texas summer day, and I climbed into a nearly cloudless sky. With flaps retracted and power reduced, I continued the smooth ascent, making sure to call the required radio checks as I changed altitude and heading on the brief route to the training area.

When I leveled off at 6,000 feet, I found myself grinning broadly. The Navy had actually given me this fantastic airplane to fly and they were paying me to do it. Was that cool or what?

On my next solo, I flew out to the training area, swiveled my head around the sky to make certain nobody was nearby, and rolled the plane inverted. Hanging from my harness, I kept my right hand on the stick and used the left to snap a picture of myself with the earth and sky reversed outside the canopy.

I figured I would complete the second half of the syllabus in the next two months, hopefully continuing to score well. Then my on-wing called me in to the squadron office. *Oh, no,* I thought, *how have I screwed up?*

But there was another student in the office, Ensign Don Shank, who I knew was a good stick.

Lieutenant Nelson, who was also the student control officer, held the key to our future training in his hands. "You guys are doing the best of anyone in your class." He looked at us both. "You're natural aviators."

Coming from him, that was a real compliment. But this wasn't just a confidence-building session. It was September 10, 1997; we were almost at the end of the fiscal year and the squadron hadn't graduated enough students yet due to its own high standards and the bad weather earlier in the summer.

"If you agree, we're going to double-pump you," Nelson said.

This meant Don Shank and I would be either flying or simming at least twice a day, six days a week, to complete over half the Primary syllabus in just two and a half weeks.

We immediately moved into Precision Aerobatics, during which I alternated among flight instructors. In three days I had learned all the barnstormers' aerobatic maneuvers that had evolved out of the elaborate dogfights of World War I. I was able to fly loops, aileron and barrel rolls, Immelmanns, and the split S, which is basically the bottom half of a loop that reverses your heading. I'd practice all of these maneuvers for hours, stringing them together into a neat sequence. Then an instructor would observe from a chase plane and put me through my paces for a score. As always, I was a little nervous being watched this way. But I kept racking up one Above after another.

I finished Primary by compressing the formation-flying curriculum (FORMs) into just three days. Flying with Don Shank as my partner, we practiced exacting maneuvers that involved linked turns, breakaways in opposite directions, followed by rendezvous on exact headings and altitudes. On FORMs we switched flying lead and wingman, first with instructors in the backseat, then observing us from chase planes as we flew solo. We flew in trail formation and close-up, wingtips only four feet apart. Because the T-34Cs were stable, extremely maneuverable little airplanes, this flying was actually much less hazardous than it first appeared.

Somehow I had managed to study and prepare for each flight and transform that preparation into solid performance. This had been both an exhausting and a satisfying experience. I alternated instructors, flying with the most experienced men in the wing. They knew Don and I were being double-pumped and made certain we were up to the effort, while never letting the training standards go slack.

I finished Primary flight training at Corpus Christi on September 30, the last day of the fiscal year. My total on the Naval Standard Scoring was fifty-five. You needed over fifty for jet training and I had made the cut by a good margin. All the hard work had paid off.

But there was a problem. The Navy's fleet of T-2 jet trainers had been grounded for three months due to flight-control problems. This meant that there was a long backlog of candidate pilots waiting in the pipeline for T-2 training, and the Navy wasn't accepting any more. No T-2 training, no fighters.

I'd known for weeks that the jet pipeline was crowded, and had learned just how bad the situation had become before I'd had to decide on accepting double pumping. If I had delayed completing Primary at Corpus, I might have escaped the jammed T-2 pipeline and made it into jets later the next fiscal year because of my grades. Instead, I had done what was best for the squadron.

Lieutenant Nelson again told me to report to office. I knew what was coming, and he didn't waste time getting to the point.

"I'm sorry, Shane," Lieutenant Nelson said. "Jets just aren't there for you. I think you knew this was coming if you completed Primary during this fiscal year. But you didn't complain a bit. I'm proud of you for sucking it up and not making a big deal about it even though you understood what it would cost you. A lot of other guys would have gotten a medical down to delay them in the curriculum, but you just kept flying day after day."

He shook his head. "I really am sorry. You worked so hard for this."

I didn't know what to say.

The door had slammed shut in my face. A future as a Navy fighter pilot was suddenly closed to me. This was the first real setback in my life. I wasn't sure how I was going to handle it. After working so hard, I had never considered this possibility.

It was not fair. Then I realized that life often was not fair. I thought about my mom and the vets back in the VA home in Norfolk. I lived in the real world, and I had to accept it. Officers served to meet "the needs of the Navy." And the Navy did not need any more fighter jocks just then.

I was twenty-three years old and had dreamed of flying since I was four. I had to get on with life and submit choices for other types of airplanes.

My first was the E-6 Mercury, a military version of the four-engine Boeing 707 that served as an airborne command post for our intercontinental missiles. Unfortunately, slots for the E-6 pipeline were also taken.

I didn't have any better luck with the E2-C2 Greyhound carrier-based turboprop plane. My final choice was the P-3C Orion maritime surveillance aircraft. It was a big four-engine turboprop sub-hunter based on the old Lockheed Electra airframe.

There were immediate openings for training in the P-3C pipeline. So, that was the direction I was heading.

Within a week of receiving the bad news about T-2 training, I got word that I would begin Advanced Training in November to enter the P-3 pipeline the next year.

My spirits rebounded. I would not be flying jets, but I would be flying. And I was determined to become one of the best P-3 pilots in the Navy.

Chapter Three

# SOUTH CHINA SEA

## 1 April 2001, 0830 Hours

We banked right to the track exactly on schedule. The sky was still cloudless, the ride smooth. I eased the four power levers with my right hand and watched the shaft horsepower and TIT drop evenly on the engines. Wendy Westbrook reached up to adjust prop synchronization and minimize vibration.

"One hundred eighty knots on the nose, Shane," Pat Honeck announced.

My airspeed indicator matched his. "Roger that, Meeso."

Regina Kauffman checked in to report that our heading to the first turning point was "right on the money" and that drift was not a problem.

I held the control yoke firmly to verify for myself that we were on the correct heading, at the briefed altitude of 22,500 feet. Everything looked good, so I re-engaged the PB-20N autopilot, and the plane settled solidly on track. Then I flipped my ICS to the loop that connected directly to Johnny Comerford.

"Hey, Ballgame, what's the situation back there?"

"Nominal, Shane," he said. "Systems are up and running. We're getting some activity back here."

By that he meant our powerful sensors were already sorting interesting emissions from the invisible electronic blizzard around us and that the in-line operators were analyzing some of these signals. Great. That was why we flew these missions.

My eye went to the fuel gauges on the center console. We were in good shape and could extend our track time if there was a good operational reason, as long as the weather en route back to Kadena did not deteriorate.

So far, this had indeed been a textbook mission. I could only hope that the Chinese didn't decide to intercept and harass us the way they had the week before.

That morning we had been headed to the southwest with a lot more time remaining on track than we had today. I'd been in the left seat, Jeff Vignery in the right. The familiar pair of Chinese People's Liberation Army/Navy F-8 II Finback fighters had appeared off the right side, as usual flying between us and their base at Lingshui on the Island of Hainan. On some intercepts during this Det, the Finbacks had just given us a quick once-over then dropped away and returned to base; on others they had stayed out there and hounded us for much longer. Now they seemed inclined to remain with us.

"They're co-altitude at three o'clock, closing," Jeff had said.

Straining forward in the left seat, I could see out the right cockpit side window. There they were, two long, sleek gray jet fighters, their modified delta wings razor-thin and their towering, swept-back vertical tail stabilizer fins massive from this angle. The Finbacks flew in their familiar clumsy loose trail, nose high as they wobbled along, trying to match our slow speed.

As we watched, the fighters slid even nearer to our right wing.

I alerted the crew to begin the standard intercept procedure. Observers had gone to their assigned windows, and Johnny had taken his clipboard to make detailed notes on the fighters' side numbers, ordnance, and behavior.

In the past, intercepts had usually lasted a couple of minutes, max. But now the Finback pilots had seemed determined to dog us. The lead fighter was mushing toward our right wingtip.

"Boy, they're close," Jeff had said. "God, they're close."

"Meeso," I'd called Pat Honeck on the intercom, "come on up here to get some pictures."

Pat had piloted S-3 Vikings off carriers and had a lot of experience formation-flying.

"Tell me how near they're getting," I had requested when he reached the flight station with his camera.

"He's probably fifty feet now," Pat had coolly announced, "maybe a couple of feet closer."

"Take some pictures of this idiot," I requested. I wanted to have photographs showing the Finback's bright red side number as evidence of this reckless airmanship.

I could see the Chinese pilot's head distinctly through the streamlined dome of the fighter's canopy. His white helmet was emblazoned with a large red star on the brow, and his dark-tinted visor was down, touching the upper curve of his oxygen mask. None of his face was visible. I might have been looking at a doll pilot in a model. Except that this was no model airplane. The Finback was huge right out there beside our wing. From my vantage point, I couldn't see the fighter's entire 70-foot length.

The Finback was certainly too damned close for safety. So I turned slowly away, easing off toward the north.

For the next forty-five minutes, the two fighters had dogged us, one on our left wing, the other on our "six," straight behind us, keeping station between our airplane and Chinese airspace.

*The hell with them,* I'd thought. *We have every right to be out here. We're in international airspace, and I'm not going to let them push us off track.*

But I kept my hands resting lightly on the yoke, ready to take over from the autopilot and bank away if a fighter approached too close. When the Finbacks finally did head back toward their home base, my neck had been knotted from the tension.

My stomach growled. It had been over six hours since I'd eaten a breakfast bar and gulped a little carton of OJ in my room at the Habu Hilton, the Bachelor Officers Quarters at Kadena. The BOQ was comfortable enough, furnished like a Holiday Inn. The sheets were clean and the water in the showers nice and hot. The "Habu" in question was a small green viper common on the island of Okinawa that the shopkeepers put into sake bottles and sold to the Marines and airmen. Pickling a poisonous snake was supposed to improve the quality of the sake.

I couldn't say; I've never been a sake drinker. My tastes had always run more toward Jack Daniel's bourbon and Coke, a deceptively sweet concoction that was a favorite among Naval Aviators. In fact, at the last crew party the previous week, some of us might have had a couple too many. No harm done; we weren't flying the next day, and I believed that a cohesive crew worked hard together and played hard together. That is an old Navy tradition that I have every intention of keeping intact.

This time of year, it was easier to both work and party at Kadena than up at Misawa because the weather in Okinawa and on our track to the south was generally better. And today was an excellent example of just how good the weather could be down here if we didn't encounter sudden pre-monsoon thunderstorms.

Johnny Comerford gave me regular updates on the progress in the back end. The way things were going, this was going to be a worthwhile mission. I felt a deep sense of satisfaction. The EP-3E crawling along at 180 knots was certainly not an F-14 Tomcat or an F/A-18 Hornet screaming through the sky at Mach 2, but we performed a vital job for the Fleet and for the country. And I was proud to be mission commander less than two years after I'd joined VQ-1 at Whidbey Island, Washington.

Those two intense years had been among the most challenging and rewarding of my life. And they reinforced my belief that hard work eventually paid off.

In October 1997, within a week of receiving the bad news about jet training, I got word that I would begin Intermediate Training at once, so that I could be ready to start Advanced in November with a new squadron that was forming.

That squadron, which would soon become VT-35, was the first to use the new twin-engine turboprop TC-12s, the military trainer version of the Beech King Air 200. The TC-12 was a lot more airplane than the Turbomentor. It had more powerful engines, pressurization, which allowed higher cruise altitude and longer legs, and carried the full range of radio navigation instruments.

But before I could qualify to begin Advanced Training, I had to complete Intermediate. Now Lieutenant Jeff Nelson showed his gratitude for my sticking to it through the double pumping and completing Primary by October 1. We began an almost daily series of long cross-country visual and instrument navigation flights, flying a T-34C all over Texas and points west. We got in a nice aerial tour of the Grand Canyon and even managed a weekend at Las Vegas. Now I was beginning to realize how much fun flying in the Navy could be. I won about fifty bucks playing blackjack, the steaks were cheap, and the casino drinks were free.

Jeff Nelson figured a way to expedite my Intermediate Training was having us park at Corpus Christi International Airport to avoid the long taxi times and traffic patterns around the Naval Air Station. This was just an extension of double pumping. We managed to complete a three-month curriculum in just twelve days. I was ready to move on to Advanced Training with VT-35, which was also based at Naval Air Station, Corpus Christi.

I was tired, but understood more firmly than ever that when you train for the demanding profession of aviation, the more intensely you flew and studied, the better you became.

The situation at the new squadron was very promising. There were only four students in the first class, and all of us had come out of Primary with jet grades. We would have our choice of twelve aircraft, which meant there would never be maintenance down. Since we would be the squadron's "laboratory rats," our instructors made it clear they needed our help in writing the syllabus for later classes.

I paired up with Ensign Kate Standifer as my plane mate. She was a natural pilot and a quick study. Our on-wing was Lieutenant Don Hyde, a laid-back former C-2 pilot who loved to fly as much as anyone I've ever met. It was Don Hyde who thought of the squadron's call sign, Sting Ray, from the aircraft's rakish T-tail. Kate and I maintained a friendly competition with the other two students, Ensigns Mark Russell and Shane Sullivan. I suppose this was inevitable when you had two Shanes.

Since we didn't have a Cockpit Procedure Trainer or a simulator for the TC-12, we spent a lot of time in the actual cockpit, with Kate or me swapping into the left seat, while Lieutenant Hyde flew.

The TC-12 is a multiengine aircraft, which meant we had twice as many engine instruments and controls to master. The plane also employed a crew of two. This entailed vital cockpit coordination, forming a close-knit professional partnership and trusting the other aviator to backstop you just as you filled the same role for them.

For example, one of the first big challenges was properly shutting down an engine in an emergency and continuing the flight on a single engine. Training for this situation ranged from a reasonably benign shutdown of an overheating power plant while at cruising altitude to losing an engine on final approach or soon after rotation on takeoff. Since we didn't have simulators, we had to practice these procedures on the ground with our instructors, all the while keeping careful notes on the process and debriefing during long sessions so that a detailed and useful training syllabus could be written for the students who would follow us.

I remember one morning in February when I had to fly an engine emergency with Lieutenant Chris Burkett, a friendly former EP-3E VQ-1 pilot from Georgia. I didn't know exactly when he was going to pop the emergency on me, so I had to be prepared for anything. I was flying the takeoff from the left seat. It was nice, cool weather, with clouds and showers forecast for the afternoon. I got my takeoff clearance from the tower, turned onto the active runway, and started the roll. We had briefed rotate at 110 knots and Burkett let the flight proceed through my call for "gear up" and the procedural departure turn to the left at 1,000 feet.

Then he announced, "Engine fire on number one."

Every sense I'd developed so far as a T-34C pilot warned me to leave that engine alone. Power meant lift; lift meant climb; climb meant life. But fire could kill you very quickly. I followed my briefed procedures.

"Emergency shutdown checklist," I said.

Lieutenant Burkett called out each item, and I responded, "Concur."

I kept the yoke at the same easy angle of climb and evenly pulled back the number one power lever to Flight Idle, compensating for the lost propeller torque with the rudder. I called for Lieutenant Burkett to raise the flaps, and then added, "Fire bottle number one." I hoped my voice was not too shaky. We continued to climb. "Feather number one."

Burkett's hands flew across the cockpit console. "Concur. Flaps up," he responded. "Concur. Simulated fire bottle." We weren't going to mess up a perfectly good turbine with fire extinguisher chemicals.

I held the climb out, simulating a call to the tower to declare an emergency. Although I had studied the aircraft manual, I guess I hadn't really been prepared for the fact that this aircraft could climb so well on a single engine. When we reached 2,500 feet, I leveled off and we restarted number one. Later that day, Kate flew the same emergency.

That's the way the training went over the coming months. Each week I learned some new and complex piece of disciplined airmanship. It was like building on a solid foundation.

But we also had fun. Because of the plane's range, we could fly over to New Orleans or up to Ellington Air Force Base outside NASA's Johnson Space Flight Center in Houston. There was plenty of empty air around South Texas, and we could really put the TC-12 through its high-performance paces. Now on each flight we wove in radio-instrument navigation problems, flying compass radials from VOR or TACAN ground beacons and monitoring our progress with DME (Distance Measuring Equipment).

Kate and I flew together a lot, one of us responsible for navigation, while the other handled the aircraft. On some days we'd change with the other two students, and I'd find myself in a cockpit with two Shanes. Then Mark Russell and I flew all the way up to Buckley Air National Guard base in Denver. We were becoming real pilots.

Because we were working so hard on the syllabus, it was understood we'd get our choice of assignments when we graduated from Advanced and earned our wings. One day I flew up to Ellington with Lieutenant Chris Burkett and we had lunch at the Hooters near NASA. The Buffalo wings were great, but we had to settle for iced tea instead of Dos Equis longnecks because we were flying.

"You know, Shane," he said, "the EP-3E is a lot more interesting a mission than sub-hunting in a P-3. You might not get to bomb anybody or shoot any missiles, but you fly in some pretty interesting parts of the world."

Chris had been with VQ-1 when the squadron had moved from Guam to Whidbey Island in 1994. And he knew the airplane and the mission very well. The more he described them to me, the more convinced I became that I wanted to go the VQ route. What made this choice especially attractive was that one of the EP-3E squadrons, VQ-2, was based at the Naval Air Station in Rota, Spain, and staged out of Crete, flying missions over the Balkans. As far as SIGINT went, that was the big leagues. Besides, I had studied Spanish and liked the language. A three-year tour in Rota would be great.

But first I had to type-qualify on the TC-12, a requirement for receiving my wings. One evening in May, when we had about 150 hours flying time in the aircraft, Kate and I completed our solo night flights. We both passed that demanding hurdle. The final requirement would be the full-up check ride in which an instructor would put us through the wringer, testing every skill we'd learned.

As I approached this pivotal point, a crisis loomed in my personal life that I could no longer ignore. In college, I'd been going steady with a girl named Jen, who attended school in Colorado. We'd become engaged when I was in flight school; she moved to Corpus Christi. And we planned a big wedding in Colorado one week after I received my wings. But the more I thought about getting married, the more I realized I wasn't ready. By definition, my life as a young Navy pilot was going to be unstable. If I went the VQ route in EP-3Es, I would be on long detachments away from my squadron's home base. That was not a good way for newlyweds to start their marriage. There was no way around the problem. The weekend before my final check ride, I sat down with Jen and told her my thoughts.

"I do not plan to get divorced," I said as honestly as I could. My parents had been divorced, and I'd seen how much pain it could cause in other families.

At first Jen was stunned, but she understood my reasoning.

We spent the next day printing out and addressing "don't come to the wedding" requests to the long list of friends and relatives on the invitation list.

That Wednesday, I was sitting in the left seat of a TC-12, flying patterns above the Corpus runway, as the instructor threw one order after another at me. When I banked left, I could look straight down at my apartment complex. There was Jen loading suitcases into her car. She was moving out of my life.

I kept flying my check ride to the best of my ability. In the Navy they teach us to compartmentalize our feelings. If you can't leave your problems at home, don't fly that day. You'll just kill yourself or somebody else. It's a lesson I have had to learn.

On May 22, 1998, I received my Wings of Gold as a qualified Navy pilot. I was one of the few ensigns winged that warm spring afternoon. Most of the officers around me had been in training long enough to have been promoted to Lieutenant (jg). It had been exactly one year to the day that I flew my first FAM with the stern Lieutenant Jeff Nelson. I realized I was moving on a fast track.

My family joined the ceremony and we continued the party in Mexico at Nuevo Laredo in the Rio Grande valley. I learned a little bit about tequila and hangovers that weekend.

Then I got my orders to report to VQ-2 in Rota, Spain, after I completed P-3 training in Jacksonville, Florida. This was fantastic news. I took great pride in the wings I now wore on the left breast of my uniform and flight suit. I was a member of an elite group.

Before heading to Jacksonville for training, I had leave coming. So I flew out to Naval Air Station, Oceana, near Norfolk, Virginia, to visit my buddy from NROTC, Lance Wiese, who was now an F-14 Tomcat pilot. Oceana had one of the last great rocking fighter jocks' officers clubs in the Navy with both a disk jockey and a live band. Following tradition, I had to get my wings baptized by Lance and his buddies. I'm glad Lance's future wife, Tracy, was there to drive us rowdy warriors home that night! There was a great bond of comradeship, even though I was a newly winged ensign who had just been type-rated in a TC-12. Obviously I wasn't a Top Gun, but the guys accepted me, although they couldn't resist commenting on the TC-12 Stingray patch I wore on my flight suit.

"Hey, Shane," one of Lance's buddies joked, pointing at the airplane on the embroidered patch, "since when has the Navy hired commuter pilots?"

It was all in good fun. I'd explained the intense training schedule and how we'd written the syllabus for the course as we flew. Even these jet guys seemed impressed.

While I was at Oceana on that trip, I searched the classified ads for a good used Harley Davidson. I had wanted a Hog ever since I'd ridden my first Yamaha dirt bike as a kid, and even the car crash hadn't changed my mind. And there was exactly the Harley hog I wanted, a 1996 Harley Heritage Softtail Nostalgia. The bike belonged to a judge who must have decided it was too much of a machine for him. It sure wasn't too much for me.

I headed off cross-country alone on my new Hog, often leaving the Interstate just to enjoy the country roads. It took me four days to reach my dad and his wife Julie's home in South Dakota. Then I continued on to Nebraska to see my family there. I had worked hard for six years and this was a well-earned vacation.

I reported to Naval Air Station, Jacksonville, Florida, "Jax," to begin P-3 training in September 1998. When training began, I ran into friends from flight school and NROTC. I also met a guy who is now one of my best friends, Lieutenant (jg) Paul Crawford. He introduced me to a wide circle of his friends in my new training squadron, VP-30. Paul and I had been born on the same day, June 21, 1974, so we hit it off from the get-go. Even though I was living in the BOQ, the guys invited me to their condos and apartments for weekend pool parties and barbecues. It was definitely great to be a bachelor.

But during the week we had to dig in and work.

The morning we were introduced to the actual airplane, I began to understand that flying the P-3 and its EP-3E electronic reconnaissance variant would be as different as flying the TC-12 had been compared to the little T-34C. Everything about the P-3 was big. The wingspan was just under 100 feet, and the overall aircraft length was almost 105 feet. The tail stood as high as a three-story building.

As the lieutenant instructor walked us around the hangar, rattling off statistics such as maximum gross takeoff weight of 70 tons, I recognized this was an aircraft finely balanced between the power of its four big Allison turboprop engines and the heavy payload of fuel, ordnance, or electronic equipment. And the human factors of the flight station crew also had to be finely balanced, with the workload spread among the rotating shifts of three pilots, two flight engineers, and a navigator.

In ground school, I entered a bewildering new realm of aviation. The course was rigorous because the airplane is one of the most systems-intensive in the American military inventory. Non-aviators don't usually realize that a propeller engine, with its hydraulic, electrical, and de-icing systems, is far more complex than an airliner's fanjet. When you have an aircraft with four turboprop engines, that complexity is multiplied exponentially. The P-3 has miles of wiring and hydraulic plumbing, all controlled by a maze of circuit breakers, valves, and pumps. We had to memorize the function of every one of those systems in ground school, and then begin the long process of learning what malfunctions could occur and lead to an in-flight emergency.

Although I had really stretched my brain becoming type-rated in the TC-12, I now had to cram in even more obscure knowledge. Veteran flight engineers and instructor pilots taught us to spot the telltale signs of trouble in an engine by watching for brief, minor fluctuations in turbine inlet temperature and prop RPM. We learned to identify transient electrical overloads and to isolate the problem through a maze of circuit breakers. We spent hour after hour, day after day, in parked aircraft or ground simulators, absorbing this blizzard of information.

P-3 training at Jacksonville was intensive, but much of it was spent on the ground. In fact, I got only eleven training flights in the aircraft while I was there. But those flights, like the rest of the training, were long, demanding sessions. By the third flight, I was able to fly a series of touch-and-go takeoffs and landings. The next flight, the instructors began to incorporate emergency procedures for which we had trained in simulators. While flying at safe altitudes, I had to shut down a simulated overheating engine using the yellow-and-black Emergency handle, and then continue the simulation with the "failure to feather" procedure to prevent the runaway propeller from windmilling dangerously.

Many of the in-flight emergencies involved the engines. But we also had to deal with electrical failures that blacked out our radio navigation instruments as well as hydraulic problems that forced us to manually release the landing gear. Flap emergencies were one of the most demanding situations we had to deal with.

Every student pilot learns to handle fast, no-flap landings as part of the training curriculum for any aircraft. But the bigger and heavier the airplane, the more critical this problem becomes. For the P-3, normal touchdown airspeed with the flaps extended in the Landing position is around 120 knots depending on weight. A no-flap landing can be as high as 145 knots. The plane has reinforced landing gear and can handle that speed. But those extra twenty knots make a difference, and watching the runway go by you so fast takes getting used to. I've never been a timid pilot, nor have I been what's called a "floater," using up half the runway to land. So I stuck my first no-flap landing right on the hash marks.

"I guess you don't believe in screwing around," the instructor commented as I reversed prop pitch to slow the plane and we rolled out.

Just before I was scheduled to leave Jax in February 1999, my orders were changed from VQ-2 in Rota to VQ-1 in Whidbey Island. Apparently another pilot who had pull with the student control officer managed to arrange the deal. I was not a happy camper. I had volunteered to work for the system and had been double-pumped right through the second half of Primary so that I missed any chance for the jet pipeline despite the fact I had jet grades. Then I had spent the next few months as a volunteer for a brand-new TC-12 curriculum with the understanding I was earning my choice of assignment after P-3 training. Now I was going to drizzly Puget Sound instead of sunny Spain. But I soon got over it and was glad to be heading to the same squadron as Paul.

But there was definitely a saving grace in all this. About two weeks before leaving Jax, I met a beautiful girl named Roxanne Faustino, whose parents were from the Philippines and whose father had retired from the U.S. Navy. It didn't take very long for me to realize this was a young woman I wanted to spend time with. And that would be easier to arrange if I were stationed in the States than overseas.

So I left for VQ-1 feeling a lot better about the sudden change in orders.

Before reporting to the squadron, I had to complete Survival, Evasion, Resistance and Escape (SERE) School up in the mountains near

Brunswick, Maine. Anybody who's ever spent time in the Maine woods in winter realizes just how cold it can get. I had come up through Scouting in Nebraska and had done my share of winter camping. But Maine was practically the Arctic compared to the prairie around Norfolk. The instructors did teach us to survive, however. We dug sleeping caves in the snow, with a deeper central pit to collect the really cold air. We skinned rabbits and cooked them over smokeless fires of dry pine branches. We snowshoed across the bleak white mountains following compass headings through the featureless pine and birch forests, evading possible capture. It was tough. Each night as I lay shivering and hungry in my sleeping bag, I thought the same thing, *I can't believe the Navy's actually doing this to me.* Even the Marines who joined the aircrew in our class found the course tough.

One of the rougher parts of SERE School was the "Prisoner of War" exercise we had to endure. The instructors were rough, but obviously not as brutal as real enemy interrogators could be. Sleep deprivation was one of their favorite techniques. I hated going through the training, but I recognized its value. I was an American military aviator who would soon be flying in potential harm's way. I had to face the prospect that I might one day be captured. After SERE training, I felt better prepared.

Puget Sound was indeed drizzly when I checked in for duty at VQ-1 in late April 1999. But the area was beautiful, especially when the mist lifted to reveal granite bluffs covered with dark green fir and spruce standing above the glittering water of the Sound. Luckily I was arriving just as the brief sunny Pacific Northwest summer was beginning, and sunshine definitely highlighted the region's dramatic landscape.

VQ-1, the "World Watchers," is the Navy's largest operational squadron with about seventy-five officers and 350 enlisted personnel.

The squadron always has aircraft and crews deployed to two permanent detachment sites, one at Misawa on the Japanese island of Honshu (from which a plane staging out of Kadena Air Base on Okinawa flies), and the other in Bahrain on the Arabian Gulf, to support Operation Southern Watch, the no-fly zone in southern Iraq.

Flying from the western Pacific (Wespac) detachment sites, the squadron's EP-3E ARIES II covers the coast of Asia from the north Pacific all the way down to the South China Sea and, if required, into the Indian Ocean. Later, the squadron also began detachments to Manta, Ecuador, flying antinarcotics surveillance flights in South America.

On checking in, I asked the operations officer, "When am I going to get a chance to go on the road?"

He consulted his status board. I hadn't qualified as a third pilot (3-P) yet. This was a rigorously structured process that included training simulators and flights as well as written Proficiency Qualification Standards tests. To become a 3-P required passing two simulators, three instructed flights, and taking a PQS about thirty pages long. An aircraft commander had to sign off on all qualification lines prior to the simulators and instructed flights. As in earlier training, I would be tested on all the aircraft's systems, turbines, propellers, hydraulics, the whole complex web of alloy, tubes, and cable that made the airplane fly.

"It looks like September," he said.

"Can't I get out of here any sooner?" I was eager to be flying operations overseas.

There was a detachment scheduled to depart for Bahrain in one month. The crew's 3-P had a pregnant wife and had requested to be excused from the Det.

"How soon can you get qualified as a 3-P, Osborn?" the Ops officer asked.

"I think I can get qualed in a month, sir," I said, "if I can get the sim and flying time."

The officer looked at me skeptically. "Let's give it a shot," he finally said.

As events developed, I did manage to get my qualification as a 3-P in time for the detachment. This was even tougher than double pumping through Primary and charting new training territory in Advanced. But it was worth the effort. When the crew left for Bahrain, I was the 3-P.

My crew on this first Det included the EWAC Lieutenant Ed Chung, who'd been born in Taiwan, and Lieutenant Berry Vaughn, the 2-P. We passed through Rota en route and onward across the Med to Crete and the Middle East. Rota did look like a great duty station, with the beaches of Spain and the exotic towns of Morocco nearby. But the orders had not worked out, so I put my regrets behind me.

Arriving at Manama in Bahrain was a shock. Our plane landed just after sunset, but the temperature was still about 115 degrees Fahrenheit. And this was not dry desert heat either. The humidity of the surrounding Gulf wrapped around you like a wet, salty blanket. I'd heard the weather was harsh but couldn't believe just how bad conditions could be in the Persian Gulf in the summer.

One of the problems with this weather was that the sand and humidity took its toll on the aircraft. Chief Wolcott and his maintainers pulled off a feat that still amazes me. Somehow, they gave us an up airplane for *every* mission, and we flew thirty-five straight missions without a single maintenance cancellation. Working on the aircraft day and night under those conditions was brutal. But none of them complained.

In these temperatures, we had to fly shorter missions because we couldn't take off in the middle of the day with a heavy fuel load for fear of losing an engine on takeoff and not being able to climb out to a safe altitude.

But there were good things to say about Bahrain. We were billeted in condos at the Manai Plaza Hotel, a fifteen-story building that the U.S. government had leased for military personnel assigned there. Bahrain is the principal hub of Gulf Air, and their Australian and English flight attendants frequented the bars and restaurants near there. There were also a lot of great air-conditioned dance clubs nearby. When we were not flying or studying for our next PQS, we partied.

Once more, the old work-hard, play-hard tradition proved its value. Our crew often went out as a group, officers and enlisted ranks treating one another equally. We all accepted military hierarchy while on duty, but I for one understood I could learn a lot from the veteran enlisted people. One of the flight engineers, Chief Petty Officer Tom "Trixie" Dutrieux, probably taught me more about the aircraft's complex, inner-connected systems than I'd learned in several months of intense formal instruction at Jacksonville and Whidbey Island.

We supported Operation Southern Watch flying a track within the borders of Kuwait. But we were always aware that Iraqi surface-to-air missiles didn't stop at borders, and that we presented a juicy target should Saddam Hussein decide to take a shot at us. Because of the heavier fuel load required, we flew our longer reconnaissance patrols over the Gulf at night. One of our responsibilities was monitoring tankers breaking the oil-export embargo the United Nations had imposed on Iraq. And we also had to keep our SIGINT profile of Iraqi forces up-to-date while ensuring the safety of the U.S. Navy carrier battle group operating in the Gulf.

Early one morning in midsummer, we were just fifteen minutes from the end of a seven-hour mission, cruising back south toward Bahrain at 24,500 feet. I was in the left seat, fighting off the fatigue of flying through the night by making conversation with Berry and Petty Officer Wendy Westbrook, the flight engineer undergoing training.

Suddenly a yellow PRESSURE LOW flashed on the fuel panel for engine number four. Ed Chung, the EWAC, and Chief Dutrieux, the senior FE, promptly came up to the flight station to stand behind us.

We were still on track, 140 miles from Bahrain. "Maybe we should wait ten minutes," Ed suggested.

Three minutes later, the engine lost about 2,000 shaft horsepower, almost half its performance.

"No, sir," Chief Dutrieux said firmly. "We're going home now."

"Roger that," Ed said.

"Emergency shutdown, number 4," I requested, and Berry concurred.

Wendy "stroked" the number four E handle, and the outboard engine on the right wing immediately shut down with the prop neatly feathered.

About fifty miles out, we got a low fuel pressure light for the number two engine, and its shaft horsepower also began to drop. On the maintenance checks before takeoff, the ground crew had complained about water in the fuel but were confident they'd fixed the problem by draining the bottom of the tanks. Now we all suspected the same critical situation: The fuel in our tanks *was* contaminated, either with water or something else, and our engines would flame out one by one until we became an uncontrollable sixty-ton glider.

"Set Condition Five and prepare to ditch," Ed Chung announced on the PA.

In the back end, where I was now sitting, we pulled on our helmets, locked our chairs facing aft, and snugged down our harnesses hard.

Although we tried to keep our faces neutral, I could see that everybody was scared. No crew had ever ditched an EP-3E in the sea before. Everyone believed that the bulbous fiberglass radome of the Big Look radar protruding from the forward belly would plow into the water so hard, creating such instantaneous deceleration drag, the plane would pitch forward on its nose and smash upside so violently that the crew would be killed in the impact. Despite the chill air-conditioning, everyone was sweating.

But we didn't have to test the chances of surviving a ditch that morning. As the plane descended into warmer air, the power began to creep back up on engine number two. We made an uneventful landing with the number four prop still feathered. When the maintenance crews and safety inspectors investigated the incident, they found the fuel tanks almost empty and a fair amount of water in each tank that had somehow evaded the purging procedure before takeoff. Then they reached a logical conclusion: The humidity had been so extreme on takeoff, we had actually carried our own water contamination with us to high altitude in the form of saturated air. As the tanks emptied, that humidity condensed into water, which sank beneath the lighter fuel and froze in the fuel lines, first starving number four, then number two.

Only when we descended to warmer altitudes did the number-two fuel line thaw out. Chief Dutrieux had saved our lives by calling for an immediate return to base.

If I ever needed a reinforcement of the advice all junior officers are given, "Listen to your chiefs. Rely on their judgment," I got it that frightening morning over the Gulf.

When I arrived back at Whidbey Island after my first Det, my training pace was almost the opposite of what it had been when I left. All the aircraft were either deployed, "broke" with one maintenance problem or another, or grounded for modifications. I spent the next four months without a single training flight.

But in order to qualify as a 2-P, I had to fly a minimum number of training hops and pass a four-hour systems board. I wasn't going to accomplish this very soon, no matter how much study or simulator time I got in.

So I was pleased to escape from this limbo by taking my second Det in January 2000, even though I was frustrated at still being a 3-P. This was a Westpac detachment that began in Misawa and continued in Kadena. Out there, the missions were much longer than those we had flown in the Middle East. And we were usually on our own, not working with an aircraft carrier battle group as we had in the Gulf. One of the features that was great about this Det was I got to fly in a lot of different weather conditions, ranging from the northern Pacific to the South China Sea. The long flights also gave me plenty of time to hone my flying skills and mission knowledge.

No sooner had I come back to Whidbey Island in April than another unusual situation arose. There was a 2-P out on the current Westpac Det who had to come back to the States to receive an award. The squadron Ops officer knew how much I wanted to become a 2-P. But to qualify, I needed nine more instructed flights and three rigorous simulators, not to mention passing the systems board composed of an instructor pilot and two EWACs. If the aircraft and sims were free, he said, I could take a shot at it.

"How much time do I have?" I asked the Ops officer.

"About two and a half weeks," he said.

That was more structured training than I'd had in the previous ten months. But somehow I pulled it off. I sat for my 2-P systems board at six in the morning after a long instructed training flight that had ended at nine the night before. I was tired, but my mind was clear. The board members threw one curveball after another at me. I managed to pass. But I still needed my last check ride to be a qualified 2-P. And there just wasn't time.

So when I returned to the Westpac on my third Det, I was still a 3-P.

It would be only a matter of time before I qualified, however. Despite the gap in training time the previous summer, I was back on the fast track.

This was a great crew. My close friend Paul Crawford was the 2-P and our EWAC was Lieutenant Norm Maxim, without question the best aircraft commander I've ever flown with. Norm was a tall, husky Boston Irishman with unruly red hair and a sardonic sense of humor that was too subtle for most people. A former enlisted man with a down-to-earth wife named Tiffany, he had a work ethic that just wouldn't quit. Norm was one of the squadron's best instructor pilots and took it on himself to mentor me, just as he had mentored Paul Crawford.

But I soon found being mentored by Norm Maxim was not easy. The first day flying from Misawa, I was sitting in the right seat, when Norm began bombarding me with questions from the left.

"Let's say you experienced sudden but intermittent shaft horsepower drops on numbers two and three. Fuel flow looks normal. How do you proceed?"

I had no sooner answered that one when he shot back, "Split between the right and left gyros. Left displays alternating shallow banks. Right shows straight and level. It's a dark night. How do you determine which instrument is accurate?" The questions kept piling up, and I did my best to answer.

But Norm Maxim's relentless quest for excellence went beyond technical airmanship. He peppered Paul and me with tactical questions from the mission commander's notebook. We were expected to learn the force structure and current geopolitical policies of all the nations in the regions the squadron patrolled. This wasn't just some theoretical college seminar, but a practical requirement to becoming an EWAC and advancing to mission commander. At that level, an officer had to understand the exact nature of the potential enemy's military threat. If they painted your aircraft with fire-control radar, how likely were they to actually launch a SAM? Naturally, we also were expected to stay as current as possible on our government's attitude toward these countries. And we had to understand the exact nature of the intelligence we were collecting and the reasons it was needed.

"Hey, Shane," Norm told me after my first exhausting quiz in the cockpit, "I'll help you study. You need to learn this stuff, and there's a lot of it."

"There sure is," I said, beginning to grasp the extent of the knowledge needed to advance.

"Well," Norm added, "it's time to start really studying. The mission stuff is harder than aircraft systems and flying. You don't want to wait until the last minute."

The only place we could study this material was in the sensitive intelligence facility at the Det sites or Whidbey Island. This meant that each night after flying, or when the plane was down for maintenance, Paul and I headed over there and began grinding through the hundreds of detailed files on the classified computers.

The hard work paid off. I qualified as a 2-P after I returned from the Westpac in May. But Norm did not let up on me. Whenever he flew as my instructor pilot, he kept slamming me with increasingly tough questions.

"Don't forget, Shane," he'd always say, "the only way you'll make EWAC and mission commander is to study now. You don't want to wait until the last minute."

There was no arguing. I spent as much time as I could at a classified computer in the Whidbey Island intelligence facility. I enjoyed being on the fast track. A lot of people just kicked back a little and coasted between 3-P, 2-P, and EWAC. I wasn't satisfied with that. Normally, a pilot is expected to qualify as an EWAC two years after entering the squadron, provided there had been adequate training and deployment opportunity. I was determined to advance faster.

That summer, I went on my second Det to Bahrain. Norm Maxim was the EWAC, and Lieutenant Todd Lacy, the Naval Flight Officer in charge of the backend, was the mission commander.

The heat was just as bad, but now the Navy had billeted us inside its own secure compound, in poorly air-conditioned enlisted barracks. The luxury of the Manai Plaza condos was a fond memory. The aircraft was in bad shape and required four engine changes in a month. I spent six or eight hours a day, every day, in the well-air-conditioned intelligence facility. If I wasn't studying, I was either flying or being quizzed by Norm and Todd. Knowing that I could fly the airplane well, they concentrated on tactical and geopolitical knowledge.

Some of our quiz sessions got intense. They would hammer me with scenarios in which the aircraft received signals from extremely hazardous ground or air radars, indicating possibly imminent missile attack. But the degree of that threat depended on the political complexity of the potential enemy.

"What are his intentions *now*?" Todd would demand. He knew this stuff cold.

"You have to make an immediate decision, Shane," Norm added. "When you're sitting in the seat, you don't have time to look up the answer."

I spent more time hunched over the computer screen.

But the hard work did pay off. In September 2000, when I got back to Whidbey Island, I passed my EWAC board, seventeen months after reporting for duty to the squadron.

Then, still officially a 2-P, I left on my fourth Det of the year, part of the first VQ-1 crew in Manta on the coast of Ecuador. Our mission was

Signals Intelligence in support of the antinarcotics effort. Lieutenant (jg) Jeff Vignery flew 3-P on the Det as 3-P and picked up his call sign, Jefe. Our navigator was Lieutenant (jg) Regina Kauffman. And my close friend Johnny Comerford was the EVAL, working with the backenders.

It was a fascinating detachment. On the way down, the plane was grounded for maintenance in Key West, which made up for all those baking days in Bahrain when it was even too hot to swim in the pool. We were practically the only Americans when we landed in Manta. For most of us it was our first taste of the Third World with the shocking contrast between flyblown shantytowns and walled villas of the local elite.

Although some of the crew was saddened or repulsed by this level of poverty, I saw things differently. Many of the families living near the airfield might have been jammed eight or ten to a two-room shack, but they always had a bright smile for us and seemed quite content with life. Bananas, papayas, and other delicious tropical fruit grew in great abundance around their houses. There were inexpensive fish of every description on chipped concrete slabs in the open-air market. They were lucky to have the plentiful sea for a cheap source of protein at that level of poverty, which was something none of us had seen before.

I wasn't naïve enough, of course, to believe they enjoyed a perfect democracy or the rule of law that we take for granted in the United States. Poor people around the world are subjected to arbitrary cruelty from the police or local authorities. They usually have to bribe to get their children in school, which means the kids receive little or no formal education. So the wheel of poverty keeps grinding on. But the people I mingled with in Manta relied heavily on their strong family bonds. They loved their children, worked hard, hoped for a better future, and endured.

It occurred to me that all American citizens ought to spend a month or two in a place such as Manta, where the average family earns sixty dollars a month, so that they would appreciate the benefits of our democratic system and free-market economy.

We returned from South America in time for Christmas. I finished up my check ride for my EWAC designation prior to Christmas leave. Needless to say, it was a shock to go from the equator to Whidbey to the blizzards of Nebraska.

As soon as my holiday leave ended, I was back at Whidbey Island, again studying hard. Usually a pilot had to fly one Det as an EWAC before qualifying as a mission commander. I hoped to get that Det as soon as possible.

So, I was surprised and pleased in February 2001, when my squadron Executive Officer, Commander Michael Pagliarulo, called me in. "Lieutenant," he said, smiling, "you ready to take a Det out to the Westpac in March?"

"Roger that, sir."

"Well, you've got it," he said. Then he grinned and extended his hand. "And you're going to go out as the mission commander."

Chapter Four

# SOUTH CHINA SEA

## 1 April 2001, 0940 Hours

I had tried once more to grab a nap in the curtained bunk, but again had not been successful. So I fetched another bottle of water from the galley and worked my way forward, pausing along the way to the flight station to chat with the cryptologic technicians and electronic warfare operators at their on-line positions.

The airplane was flying perfectly, and all the equipment in the back-end was up and functioning. Johnny Comerford assured me that this had been a textbook mission.

"Except for no intercept," I said.

He nodded. "Well, as the man said, 'It ain't over till it's over.'"

I hoped he was wrong. The intercept we had experienced the previous week had been a nasty experience.

Back in the flight station on this April Fool's Sunday, I swapped out for Pat Honeck in the right seat so that he could use the head. Wendy Westbrook had taken over for Senior Mellos, who was back in the galley, preparing rice in the electric cooker. Senior Mellos liked his rice drenched in enough hot sauce to peel paint. I usually stuck to the in-flight box lunches prepared for us at Kadena, but would occasionally eat some of Senior's volcanic cuisine.

I pulled on my headset, studied the instrument panel, and switched around the ICS nets to pick up the crew's professional exchanges just in time to hear Lieutenant (jg) Dick Payne send the regular position report to our chain of command. We were on the long final leg of today's track, heading southwest away from the Asian coast. Regina checked in over the intercom with a position update. Currently, we were seventy nautical miles south-southeast of Hainan Island.

"We've got about ten more minutes on track before we need to head home to make land time," I told Jefe and Wendy.

It was exactly 0955 hours Okinawa time. Our current Estimated Time of Arrival (ETA) back at Kadena was 1356 hours, giving us a mission of slightly less than the nine hours we'd briefed. Even the winds had been favorable today.

As we lumbered through the tropic sky, the sun climbed higher and swung in front of us. Both Jefe and I adjusted our green plastic windscreen sun visors. Jefe also cleaned and replaced his pilot's sunglasses. I never wore glasses of any kind because I couldn't stand to have anything near my eyes after the car accident in high school.

Scanning out the right cockpit window, I suddenly saw a familiar pair of Chinese Navy F-8 Finback II fighters, about half a mile out, climbing to our altitude. The People's Liberation Army/Navy had not forgotten us after all this morning. From the right seat, my view was much better than it had been during the previous intercept. The fighters were still in loose trail formation. But the lead pilot had misjudged the relative speeds and overshot us by a good quarter mile. He had to haul back on his stick to bleed off airspeed by jerking the dark bullet-shaped nose of the Finback into a high Angle of Attack. But the Finbacks were far enough away to not cause us concern. That was the kind of sloppy maneuver that would have earned you a "Down" from Lieutenant Jeff Nelson in Primary, I thought.

Once more I alerted the crew, and the observers took their stations at windows.

"Three o'clock, level," the starboard aft observer reported. "Their trail is about two hundred feet lower than the lead."

So far, the fighters had shown no sign of closing the gap.

"I've got the controls," I told Jefe, beginning a shallow five-degree bank away from the fighters, using the autopilot for the gradual turn. I had no doubt they would follow their typical pattern and stay on the island side, shadowing us before falling back to return to Hainan while we moved on to our northeast heading toward Okinawa.

If we hadn't been at the end of our mission, I would have completed the turn back and continued to fly the track line.

Now something strange happened. The port aft observer checked in. "I've got the two fighters at about our seven o'clock position low and they're closing."

Instead of drifting back to return to Hainan, the Finbacks had shifted their position to our left side and were approaching from the rear.

"They'll probably just take a little look at us," I reassured the crew. "Then they'll peel off to go home themselves because we're headed away from them."

Holding the shallow bank with the autopilot, I watched my compass swing slowly to 070, northeast, our homebound heading. The autopilot had kept the plane rock solid the entire time. But once more with fighters nearby, I rested my hands on the yoke, ready to take control if necessary. Pat Honeck and Senior Mellos had come up to stand behind us in the flight station.

The observers on the port side kept up a steady stream of reports. Lieutenant Marcia Sonon, the special evaluator, was crouching beside Johnny Comerford at the little round window of the port-side overwing hatch. "He's closing to three o'clock," she said, her tone flat and professional. "He's definitely armed. I can see missiles on his wings. He's got his oxygen mask on."

Suddenly her voice grew tense. "He's getting really close. Fifty feet. Now he's about forty feet. Oh, my God, he's coming closer. Right now he's about *ten feet* off our wing. He's making some kind of hand gesture to us, but I don't understand what he's trying to do."

Jefe and I exchanged an alarmed glance. From the right seat, I could not see the Chinese pilot's gesture. But Marcia was a trained observer and would not exaggerate.

I eased the four power levers slightly forward to increase our airspeed to 190 knots. I didn't want that guy going into a stall this close to our wing. Hell, the Finback II had a landing speed of 156 knots at sea level, so it was amazing he could hold station with us as well. He was in the thin air at this altitude. All we could do was monitor the autopilot and sweat. My T-shirt beneath the open collar of my flight suit was already damp despite the fact I had my air-conditioning vent cranked up high.

Pat Honeck leaned toward the left side cockpit window. "He's not very stable out there, Shane, and he's getting very close."

Jefe confirmed, "Yeah. He's close, really close." He looked scared. I know I must have, too.

"I'm already turned away," I snapped. Then I added in a more even tone to ratchet down the tension level, "There's no place for me to go."

Just to reassure myself, I studied the HSI compass. Our heading was an unwavering 070 degrees.

"Okay," Jeff Vignery announced, the relief in his voice obvious. "He's dropping off some."

"Now he's back toward our seven or eight o'clock position," Marcia confirmed.

That had been a bizarre, and frightening, experience. Where the hell had that guy wanted us to go? We were in international airspace, turned away from China, and I couldn't maneuver the huge aircraft with him so close. All at once I had a weird thought: *This is going to be a long mission report.* I would have to write up the dangerous intercept in detail because the Chinese pilot came so close and acted so aggressively, even while we were headed home. And my report was bound to cause a big "spool up" in the chain of command.

But the intercept had not ended.

"Port aft," the observer reported. "Here he comes again. He's closing, he's closing fast."

Pat Honeck was still at the window. "Jesus," he gasped. "He almost hit us."

I felt an icy hot flood of adrenaline. "What do you mean?"

Pat pointed wordlessly. The fighter's fuselage was just below our wing, but our two cockpits were side by side at the same level. Even from the right seat, I could see the fighter's nose was drifting in front of and inside the span of the left wing, only feet from the slashing disks of the four-bladed propellers. Now the Chinese pilot had dropped his oxygen mask. We were eye to eye and he was mouthing silent, angry words, gesturing at us with the open palm of his gloved hand as if trying to wave us away.

I was gripped by dread. How could he try to fly his plane with one hand in these conditions? The only way the pilot could control the Finback effectively was with one hand on the throttle, the other on the stick. And he wasn't doing much of a job of control because his nose was chopping up and down three to five feet at a time.

There was nothing I could do. It would be too dangerous to try to maneuver with him flying so unstably just beneath our wing tip.

Then, as quickly as he had come, the Chinese pilot dropped away again. I released a clenched breath. This *definitely* was going to be a long report.

Suddenly, Pat shouted, "Here he comes again. He's closing really fast this time."

*What the hell is this guy doing?* I wondered again.

Now Pat yelled, his words loud and indistinct. But I saw a fractured blur of what was happening at the left wing. The Finback had approached again from our left rear, closing so fast the pilot couldn't slow. Instead of dipping his nose to fly beneath our wing, he pitched up and tried to turn away to arrest his rate of closure. But he was already inside our wing. And when his nose shot up, the fighter's long fuselage rose at a steep angle toward the chopping propeller blades of our number one engine.

"God . . ." Pat Honeck shouted.

But the rest of his words disappeared in a blast of noise. A harsh, abrasive chattering filled the cockpit, as if we were speeding down a stretch of terrible washboard farm road. I saw a cloud of glittering debris blowing out in front of our left wing. The sickening chopping continued, sounding like a monster chain saw hacking metal. My hands gripped the control yoke tightly, and I felt each impact.

The crew was shouting in fear and disbelief.

Pat flinched away from the left window as the debris blew toward us. The chain-saw chopping ended, but a thick, dark shape, the front section of the Finback, flipped toward us. I had just enough time to realize that the propeller had struck the fighter where its vertical stabilizer joined the fuselage, cutting the Finback into two pieces before the fighter's nose smashed into ours.

The impact blasted like an exploding missile. Another huge, dark shape—our fiberglass nose cone—whipped up and over the windscreen and disappeared. Everyone in the flight station yelled and cringed. The first blast was instantly followed by the clap of explosive decompression and the unshielded roar of the engines as the forward pressure bulkhead was punctured.

I shuddered to see the Finback's severed fuselage continued beneath our plane, narrowly missing the right wing and the propellers of engines number three and four.

People were still screaming around me, but no one could hear.

Our nose pitched up, the left wing dropped violently, and we snap-rolled steeply left. But this was unlike any aerobatic maneuver I had flown in Primary. It was almost impossible to think with the entire plane chattering and shaking so harshly from the damaged propeller on number one and the rush of the airstream through the shattered maw of the nose.

But I immediately realized we were going into an inverted dive. That was a flight attitude even beyond the EP-3E's robust airframe. Instinctively, I swung the yoke hard right and jammed my foot on the right rudder pedal. The plane kept rolling left and the nose dropped like a stone. I caught a glimpse of the gyro. The black-and-white hemispheres were vertical. We were swinging past a 90-degree bank angle, headed toward inverted flight.

*This guy just fucking killed us,* a voice sounded in my head as I fought the controls and an overwhelming nauseating feeling of impending death sank into my stomach.

The airstream was screaming through the flight station as the speed increased.

"Get control of it!" I heard Senior Mellos bellow hoarsely from somewhere behind me. "Get control of it!"

I had the yoke turned vertical, full right, and my boot was jammed down on the right rudder pedal. But still we rolled left and the dive angle steepened.

I was looking *up* at the soft blue plain of the South China Sea. Through my right overhead window, I saw the forward half of the Finback's fuselage dropping beneath us, flames and black smoke spewing from it. What looked like a grayish-white parachute drifted away. I felt weirdly detached from the scene, as if I were watching grainy old Vietnam War dogfight footage on the Discovery Channel. Then I realized our plane was falling at almost the same speed as the crippled Finback's burning wreckage. We were not flying. We were dropping out of the sky.

*I'm going to die*, I thought, still battling the controls. This was the worst nightmare I had ever had. But I recognized it was true. *We're high. This plane won't fly. We're all dead and it's going to take a while for us to hit the water.*

As our dive angle increased, so did our airspeed. The airstream was shrieking through the holes in the pressure bulkhead. Unless I managed to roll our back wings level and raise the nose, we would continue rolling into full inversion, the nose would drop to vertical, and we'd be in a spin from which no four-engine aircraft could ever recover. Already, I saw that our bank angle was around 130 degrees and the nose had dropped 50 degrees below the horizon.

But I could not just give up and surrender to the inevitable. I was responsible for twenty-three other crew members. The Navy had trained me for years to confront emergencies. I would continue to fight as long as I could.

Logical thought began slowly to overcome shock. *Okay*, part of my mind told me, *we're snapping hard left, nose way down. Keep this full right aileron and rudder. You've got to stop the roll rate. You can't let it get all the way inverted, and we're almost there.*

*But take it easy on the pitch pullback.* From the Vertical Speed Indicator, which was pegged out at its maximum rate of 6,000 feet a minute, I saw we were dropping far faster that the airframe's design limit. But this horrible dive was also taking us into thicker air. The combination of airspeed and air density might produce enough flow over the ailerons to let me bring the wings level before our airframe was torn apart. But I couldn't haul back on the control yoke to pull us out of this dive until I did have the wings level, otherwise I'd overstress the tail for sure, we'd lose the elevators, and definitely be dead.

My mouth was cotton dry from the explosive decompression and the adrenaline pumping through my body. But I ignored every discomfort as I held the yoke hard right and gazed out the windscreen at the vast blue sea above.

Very slowly, the blue field began to slide right, then seemed to gather speed. I kept my eyes outside the cockpit, following the painstaking aerobatic instructions I had received years before from Lieutenant Jeff Nelson at Corpus Christi. "If you've got daylight," he'd told me, "get your head outside the cockpit to judge your attitude." That was a lesson no good pilot ever forgets.

There was no doubt now, the wings were gradually rolling level even as the airspeed increased far beyond hazardous maximum. But I had no way to judge this because my airspeed indicator was actually slowing. I'd worry about that later. Now I had to recover from this roll.

Just as the airstream screaming through the flight station had risen to a shrill, deafening howl, the face of ocean below us stopped moving right. My eye shot to the gyro. The two hemispheres had again reversed, with white above, black below. The wings were level. But we were still nose down, dropping fast at about 15,000 feet. The collision had occurred at 22,500, which meant we had fallen almost 8,000 feet in about thirty seconds.

Again I tried to think in a precise and logical sequence. But that was hard because I realized I was able to maintain wings level only with the yoke vertical, full right, and my foot down hard on the rudder pedal. And now the sickening vibration and shudder of the number one prop threatened to shake the wing apart. Worse, I saw the propeller was still spinning at 60 percent RPM. If we didn't get that engine shut down immediately, one of the chewed-up, windmilling blades could fly off the shaft hub and rip into the fuselage just aft of the communications station, cutting control cables and hydraulic lines and killing crew members.

I faced Wendy Westbrook in the engineer's seat. "E handle number one," I yelled. She seemed dazed. I took a chance and removed my left hand from the top of the yoke just long enough to reach over and grab the number one E handle. "E handle one," I repeated, shouting even louder.

Senior Mellos and Pat Honeck were back on their feet. During the snap roll and near-inverted dive, the plane had remained under positive G forces. People had been pressed against the deck but not tossed to the overhead. Senior Mellos tapped Wendy's arm, and I knew they would get that engine shut down. But I still might not pull us out of this dive.

With my right thumb on the yoke, I flipped from the normal intercom to the PA mode. "Prepare to bail out," I ordered.

I turned to Jeff Vignery. His eyes were wide, and I'm certain mine were even wider. "Jefe," I shouted over the prop vibration and the roar of the airstream, "get out the Mayday call."

"Mayday, Mayday," Jeff called over the international emergency radio frequency of 243 MgHz. "Kilo Romeo 919. We are going down."

I switched the radar transponder from Standby to 7700 and EMERGENCY. Now we would appear as a flashing symbol on any radar screen in range.

I knew from my training that I had to keep the crew working as a team. Johnny would be pulling the backenders together, preparing for bailout. My job now was to keep this crippled plane in the air long enough for them to get their chutes on, line up, and prepare to jump. My eye again shot to the right airspeed indicator. The needle was winding counterclockwise from 150 toward 140 knots, approaching stall speed. Yet the gyro and VSI showed we were still nose down, undoubtedly dropping much faster. *What the hell is going on?*

"Jefe," I called, "what are you showing for airspeed?"

"I'm showing it pegged."

His instrument read the maximum airspeed, 450 knots, but that was also impossible. Debris must have damaged both pitot tubes that protruded into the airstream near the missing nose cone and fed air to our airspeed indicators. So I would have to fly without indicated airspeed, relying on my gut feelings and the seat of my pants to keep from stalling or overspeeding the airplane.

The shaking from number one had not diminished. The engine instruments showed the props still spinning at 30 percent RPM. I yelled to Senior, "It's failed to feather."

Senior reached over Wendy to the upper left panel and raised the clear plastic guard shield on the number one engine electric feather button. He stabbed the button, reset it, and jabbed it again. But the prop would not feather electrically. Senior ducked back behind us to check circuit breakers.

I had a sickening feeling that the collision had overstressed the propeller shaft hub and stripped its gears. The four big alloy blades were flat to the airstream, still connected to the turbine, which itself was no longer producing thrust after Wendy had pulled the E handle.

With the windmilling blades still turning the turbine at 30 percent RPM and the missing nose cone, we had maximum drag at the end of the left wing and the front of the plane. And that drag was pulling us out of the air. We were still losing altitude at 3,500 feet a minute, now approaching 10,000 feet, if I could trust *that* instrument. I had to stop this sink rate.

I needed airspeed. Again, taking the chance to lift my left hand off the chattering control yoke, I firewalled the three good engines, jamming their power levers full forward, right past Military Power.

My eye shot from the view outside the windscreen to the Vertical Speed Indicator, to the gyro, to the altimeter. For the first time, I noticed a pair of thick black cannon plugs from the now-vanished weather radar, still attached to their insulated cables, wildly slapping my windscreen. That was the least of my worries.

Now I eased the control yoke toward my chest, terribly aware of the handgrips' weird vertical angle. I had to pull us out slowly and steadily. The EP-3E airframe is limited to a stress load of 3.5 Gs before it suffers structural damage. We'd already exceeded that load. And I had no idea what debris might have struck the elevators in the tail. I did not want to rip off a stabilizer after surviving this long.

The VSI needle crawled upward. At an altitude of about 8,000 feet, the instruments matched what I could see out the windscreen: Miraculously, we were flying straight and level. But if anything, the jolting chatter of number one's damaged prop was even worse. Holding the control yoke turned full right, with my leg thrust down on the right rudder pedal just to keep the nose and wings level, was so discouraging. *How long can I keep us in the sky?*

I turned to Pat Honeck and Senior Mellos in desperation. Pat was snapping the chest buckle of his parachute harness. "Do you want to bail out or ditch this thing?" I asked.

"Will it fly?" Pat said.

"I'm not sure," I answered. "I think so."

"Let's try landing it," Pat shouted in my ear.

He was no doubt thinking the same thing I was. There was only one way to bail out of an EP-3E: through the main door at the portside aft. But no P-3 aircrew of any country in the world had ever bailed out of the aircraft. Even Navy SEALs, freefall experts to a man, found the P-3 the roughest plane they had ever jumped. Because of the low wings, the only way the crew could exit without hitting the tail was to pull back the power on number two, the second engine in from the left, to ease the prop blast. If I attempted that with the runaway number one prop, we would roll uncontrollably again. And I didn't have enough altitude to recover.

Bailout posed another hazard. Even if most of the crew safely exited the plane at this speed and pulled their parachute ripcords, they would be strung out over miles of the South China Sea, floating in their LPUs among the sharks until rescued. Johnny Comerford could dump the big life-raft packs before he jumped, but the rafts would still fall far short of the crew members. And there were no American Search and Rescue forces in the immediate area.

Ditching was also virtual suicide. Even if the plane weren't so horribly ripped up, I'd learned in simulators that landing on the water with the bulbous "M&M" of the Big Look Radar protruding from the belly required exacting airmanship, using Landing Flaps, and carefully nursing the airspeed into a nose-high near stall before letting the tail slice into the water to slow the contact with the ocean. But I sure couldn't do much fancy flying when I had to keep in full right aileron and hard right rudder inputs just to hold the wings level. And I certainly couldn't use the engines in any delicate maneuver. We needed this amount of power just to maintain our airspeed . . . whatever *that* was.

Then, of course, there was the question of flaps. That Finback had showered the bottom of the left wing with sharp debris and I had felt his tail smack the left aileron and undercarriage before falling away. The left flap was probably damaged. If I tried dropping flaps and only one extended, the plane would again become unflyable. Hell, changing flight characteristics in any way would probably send us out of control. However we set this airplane down—on water or land—it was going to have to be a no-flapper.

Then I visualized the gaping hole in the nose where the streamlined cone had shielded the weather radar. Above a certain speed, water was as hard as concrete. Even if we somehow managed to belly-in to the ocean without immediately flipping over, when our hacked-off snout hit the sea at 130 knots, that concave gap in front of the flight station would act as a scoop, pitching us forward with such violence that most of the crew would be killed.

But we had to prepare anyway. I flipped the intercom to Navigator and called Regina. "Get us to the nearest landing strip, Reggie." I knew it had to be on Hainan Island, probably Lingshui, where that Finback had come from.

So either we were going to land on a runway today or find a way to bail out.

"We've got cherry lights," Wendy suddenly announced, her voice thick.

I saw the three red overheat lights on the engine gauges. We had flown too long with the engines firewalled. If we didn't ease back on the power quickly, we might blow a turbine. And this plane would definitely not be flyable if we had to shut down another engine, especially if it was number two on the left wing.

I flipped the switch on the inertial navigation repeater located on the center console to display my ground speed. This speed did not take into account tailwinds or headwinds at this altitude, but the instrument's digits gave me my only estimate on what my airspeed was. The inertial gave a ground speed of 320 knots. We were certainly safe from a stall.

"Military Power," I told Wendy.

She pulled the power levers back and the cherry lights blinked out one by one. Normally, we could fly at this power setting for only thirty minutes without causing turbine wear. But this was not a normal day, and I cared about airspeed, not wear and tear on the turbines.

"Activate the Emergency Destruction plan," I called over the PA, "and prepare to ditch."

"Jefe," I called, "get me the ditching checklist." Most of the items he read did not apply to this real-world emergency. There was no need to

"Easy Rider"—my first motorcycle at age twelve.

Graduating from the University of Nebraska—on to the Navy!

My first flight instructor, Lieutenant Jeff Nelson.

At Survival, Evasion, Resistance and Escape (SERE) School
in Brunswick, Maine.

My first crew in Bahrain, 1999.

On the flight deck of the U.S.S. *Kitty Hawk*—part of a thirty-day Midshipmen First Class cruise performing flight operations in the Pacific.

At Primary Flight Training with a T-34C, the first aircraft I flew in the Navy.

A Navy photo of the EP-3E ARIES II (Airborne Reconnaissance Integrated Electronic System II), the exact type of aircraft I piloted in China.

The Chinese F-8 jet flying dangerously close to our plane, moments before collision.

Our damaged EP-3E plane at the Lingsui military airfield on the Chinese island of Hainan.

The damaged propeller roots and hub from the number-one engine on the EP-3E. Debris from the Chinese fighter tore into the roots, almost severing them. If one of these huge prop blades had torn loose, it could have slashed into our plane and destroyed it.

Inspectors examine the severed nose of the plane.

We are greeted by military officials at Hickam Air Force Base, Hawaii, after we were held on Hainan Island for eleven days.

My whole family lined up to greet me at Whidbey Island Naval Air Station in Washington on April 14, 2001, two days after arriving in Hawaii.

The awards ceremony at Andrews Air Force Base, on May 18, 2001, where I received the Distinguished Flying Cross and the Meritorious Service Medal.

Returning a salute from Chairman of the Joint Chiefs of Staff
General Henry H. Shelton at Andrews Air Force Base.
Secretary of Defense Donald H. Rumsfeld looks on.

With President George W. Bush in the Oval Office.

My EP-3 crew (*from left to right*): Lt. Shane Osborn, Lt. Pat Honeck, Sr. Chief Nick Mellos, Lt. (jg) John Comerford, PO2 John Hanser, Lt. (jg) Jeff Vignery, PO2 Ramon Mercado, PO1 Shawn Coursen, PO2 Brad Funk, PO2 Rodney Young, Lt. (jg) Rick Payne, Lt. (jg) Regina Kauffman, Sgn. Brad Borland, Lt. Marcia Sonon, Ens. Rich Bensing, PO3 Steven Blocher, PO2 Ken Richter, AN Curtis Towne, PO2 Wendy Westbrook, Sgt. Mitchell Pray, Sn. Jeremy Crandall, PO2 David Cecka, and PO2 Scott Guidry. Not pictured: PO2 Joe Edmunds.

depressurize with these gaping holes in the pressure bulkhead. And we wouldn't be using any flaps. With two inoperable airspeed indicators, ditching speed would have to be a guestimate. The only items that really applied were setting Condition Five at the appropriate time with the backenders' seats locked facing aft and the whole crew's harnesses locked down tight. At least we were making an effort. But I knew that effort would not change what we faced.

I could picture the frantic activity in the back end. Operators and technicians would snatch up their binders of classified information, and Johnny would collect them in the crypto boxes he had used to carry the material aboard the aircraft that morning. Johnny would also supervise erasing any classified digital information. Then he would continue the Emergency Destruction checklist by smashing laptop computers with a fire ax. The final item on the checklist was dumping the boxes of classified material and smashed computers out the starboard over-wing hatch. We were well out to sea. The boxes would sink and the paper on which the classified information was written would quickly dissolve.

Pat and Jeff were obviously concerned about my "prepare to ditch" order. I hadn't changed my mind about not ditching, but I had to keep all options open. In the inverted dive after the collision, we probably reached speeds over 400 knots and had definitely been battered by G forces well above the plane's design limits. Hell, the rugged P-3 airframe had *not* been designed for a 130-degree bank angle. It was possible the plane had suffered major structural damage. The constantly hammering number one prop might shake loose, causing additional damage that could send us into the ocean immediately, in a semi-controlled ditching attempt or an out-of-control crash.

Maybe a weakened hydraulic line—possibly up in the nose-wheel steering—would split and we'd lose all our hydraulic fluid and, with it, hydraulic boost, the airplane's equivalent of a car's power steering. Then I could never keep control with the savage drag on the left wing and the need to maintain a high airspeed to prevent stalling. If that catastrophic malfunction hit us, I'd have to try bailing out the crew with this power setting on the engines, then ditch the plane myself. Some of the crew might make it. But I knew I didn't stand much chance.

Giving up at this point was just not an option. The Navy did not train its pilots to quit. I would try to hold the crew together to fly this airplane as long as that was physically possible.

But I knew our only real hope of surviving was to make an emergency landing at an airfield. The Emergency Destruction plan was well under way far out to sea. The country was not at war. There was nothing to be gained from us dying if we could save the airplane and the crew.

"Regina," I called again, "give me a heading, I need a heading to land *now*."

"About two nine zero," she answered.

We had come out of the dive flying almost due west, 270 degrees. All I had to do was ease the nose right 20 degrees to be on a rough course to Lingshui military airfield on the Chinese island of Hainan. That's where we were going.

"Lingshui, Lingshui," I called on the international emergency frequency. "Mayday. Kilo Romeo 919. We are a severely damaged aircraft. We are about seventy nautical miles southeast of Hainan. Request permission to make emergency landing."

Through the engine roar and screaming airstream whipping around my legs and chest I heard no reply. I frantically repeated the call, making sure I was transmitting on the correct radio and frequency.

Again I heard no reply.

My arms and shoulders were cramping with the strain of holding the chattering yoke at this angle. "Jefe, I need to get in the other seat."

I nodded to Pat and down to the controls, indicating I wanted him to swap out for me while I pulled on my chute and traded places with Jeff in the left seat.

Senior Mellos emerged from the alternate circuit breaker panel behind my seat. He was not wearing his parachute. *Probably thinks he's going to ride this thing down with me.* "Senior," I ordered, "get your fucking chute on."

"Roger that, sir."

As Johnny Comerford later told me, just before the collision he had been at his assigned portside intercept window, writing on his clipboard a description of the Finbacks' actions. On the lead fighter's third pass, Johnny nervously watched the Finback mush in to float under our left wing tip, nose high, precariously holding that position on engine thrust.

Johnny was shocked. *We've never had a fighter get under the wing and slide in front like this*, he thought.

Hunkered down on his haunches, Johnny was listening to the strained reports of the other observers when the Finback angled up and collided with the number one propeller. The chopping hack of the prop striking the fighter's fuselage was horrible. But that sound disappeared in the explosive impact from the nose, instantly followed by the decompression blast and the roar of the airstream.

Our plane snap-rolled left with such force that Johnny and everyone else not strapped down were pinned to the deck, unable to stand. When the wings finally went level, he grabbed the arm of a seat and dragged himself upright, then worked his way aft up the sloping deck to reach his own seat, where he pulled on his SV-2 survival gear.

The plane was still nose down, the engines screaming.

"Prepare to bail out," my strained voice blared over the public address system.

Fighting the angle of the deck, Johnny went farther aft to supervise the distribution of the parachutes. Despite his shock and fear, he was proud to see the crew behaving exactly as they had been trained. The operators were lining up to pass the parachutes forward, each to the proper person. This was essential because the individual harnesses had already been adjusted for each crew member's size.

"Position fifteen," a crewmember called, passing a bulky green backpack chute to the person ahead, who repeated the word, "Fifteen."

The process continued smoothly. Nobody panicked. After the initial shock, nobody yelled in terror.

Johnny pulled on his own chute, helmet, and gloves and sat at the jump seat beside the main cabin door. He plugged his helmet mike jack into the intercom and tried to contact me on the proper loop. "Flight, SEVAL. We're preparing to bail out." There was no response. He called again, unsuccessfully. The crew had formed a line, holding on to the overhead grab rail, and were checking each other's chutes and LPUs, almost as if this were a training exercise.

Johnny pointed to Sergeant Richard Pray, the only Marine on board. "You're my runner," he shouted. "Get up to the flight station and tell them we're ready to bail out."

"Aye, sir," Pray repeated formally. "We're ready to bail out."

Then the PA system sounded again, "Activate the Emergency Destruction plan and prepare to ditch," I called.

In a matter of seconds, the crew had left the grab rail and were snatching up their classified binders. Johnny called for the fire ax. He smashed down viciously through the keyboards of the classified laptop computers that the operators collected, making sure to mangle the hard drives below. He went forward to the starboard over-wing hatch and loaded the jettison boxes with the binders, counting to verify each box was full and that nothing had been left behind.

Johnny was ready to open the hatch and complete the Emergency Destruction checklist.

"You've got the controls, Jefe," I yelled, watching to make sure he could hold the yoke and rudder hard right. Then I popped my harness and jumped out of the right seat. Pat Honeck was back in that seat with the harness locked moments later.

"Oh, my God," Pat moaned as he took the controls from Jeff and felt the strain for the first time.

"You got it?" I couldn't let the plane roll again.

"Yeah," Pat answered with grim determination.

Johnny Comerford suddenly appeared in the narrow aisle behind the flight station. He wore his helmet, parachute, and Nomex gloves. His face was dripping sweat.

He helped me into my parachute, but my hands were shaking too badly to snap the lower buckles. "Joh—" I tried to speak, tongue stuck to my palate. My mouth was parched from the explosive decompression and the adrenaline still surging through me. I grabbed my water bottle and chugged down half. "Johnny, help me here."

He bent to snap the harness buckles. I didn't want to wear the chute because I knew there was no way I could leave the flight station after the rest of the crew had jumped and successfully work my way aft to jump myself. The autopilot could never hold the plane, and the moment I released the controls, the plane would snap-roll left again, all the way into a full inverted spin, and I'd be plastered to the deck like a squashed bug on the windshield of a Nebraska farm truck. But I had just ordered Senior to put on his chute, so it would be bad for morale if I didn't do the same. We were still U.S. Navy aviators, and I was determined to save as many of my crew as I could.

"I'm ready to pop the hatch," Johnny shouted over the surrounding roar. Neither of us was wearing a headset.

I nodded vigorously and yelled back, "Do what you have to do. I don't care when you pop it. We're heading to Lingshui."

I finished my water and swapped with Jeff to fly from the left seat. Senior Mellos had already swapped out with Wendy in the flight engineer's seat. Out the left window, I now could clearly see the damage to the number one prop tips, even though the blades were windmilling at 30 percent RPM. In the blurred disk, it looked like about a foot had been chewed off two of them. No wonder this plane was shaking itself apart. We had to get on the ground before that prop threw a blade like a 150-pound sword through our fuselage.

I pulled on the headset, took over the jolting yoke, and looked to my right. Would Senior and Pat obey an order to bail out if I gave it and stayed behind to fly the plane alone? I hoped I wouldn't have to face that question.

Pat was calling Lingshui directly on the emergency frequency. I felt the pressure change and I knew Johnny had pulled open the starboard over-wing hatch to dump classified material. A few minutes later, I felt the hatch snap shut.

Our altimeter had suddenly stopped functioning, with the needle wobbling well above and below 8,000 feet, even though the horizon and our other instruments told us we were flying level and not sinking. I wasn't too worried because we still had our radar altimeter that would begin giving a precise readout once we reached 5,000 feet.

Hainan had a tall central mountain that should be visible at this altitude and distance. But there was just too much haze to provide the visibility we needed. I wanted to slow down and gradually bleed off some altitude as we approached the island because I wasn't sure of our final heading to Lingshui.

"Reggie," I called, "where's this field?"

"Hold two nine zero until we can see the island," she answered.

I decided to start our descent then. The engines were still near maximum power, 1050 TIT. I took the three power levers in my right hand as I held hard to the shaking yoke with my left and pulled off about 1000 shaft horsepower. The horizon tilted viciously, and the left wing and nose dropped sharply. Shocked by the sudden violent altitude loss, I shoved the power levers forward again and felt the airspeed overcome drag as the props bit. The descent would have to be a lot more gradual at high airspeeds if we were going to keep this plane under control.

The hatch popped again.

In the backend, Johnny Comerford later explained, he leaned into the roaring slipstream, one hand gripping the hatch frame, the other a smashed laptop. He wanted to toss the computer well clear of the wing and thrust his torso even farther out the open hatch. But the deafening prop blast from the number three engine threatened to suck him out of the plane.

"Hang on to my chute," he yelled to the two crewmen behind him.

They took a tight grip on his harness and he pushed back out. One by one, he threw the flat black laptops into the blasting air. They blew away like dead leaves in a gale.

The Emergency Destruction checklist was complete. Johnny lifted the hatch and latched it down in its frame.

• • •

"Set Condition Five," I told Pat. "We're going to try to land."

Pat punched the PA button to relay the order. "Set Condition Five."

I pulled back on the power levers in tiny increments until we could hold a slow rate of descent. My arms were exhausted, so Pat and I took turns at the controls, each managing to hold wings level and keep us on our heading for three or four minutes at a time.

As we reached the denser, more humid air around 5,000 feet, we found we could carefully turn the yoke left from the vertical position back to about 75 degrees, making it somewhat easier to fly the airplane. But with the incessant jackhammer vibration, this was still tiring work.

Finally, the pale green hump of the island emerged through the solid haze ahead. I had the controls and Pat made the radio calls.

"Lingshui," he said twice, speaking clearly and slowly. "Kilo Romeo nine one nine declaring an emergency. We are a severely damaged aircraft approximately five miles south of Hainan. We need to land immediately."

There was no reply.

As we neared the coast, the white jumble of a small city shimmered through the haze directly ahead of us. Navy pilots do not fly crippled airplanes over cities. I added a little power and turned into a shallow right bank, away from the island. We would have to fly a circle out here until we had a better idea of where the Lingshui airfield was located. Unfortunately, we did not carry detailed approach plate charts that would have given us exact headings, runway lengths, the radio frequency of the tower and their navigation aids, and the altitude of ground obstacles we had to clear before landing. We always flew with approach plates for the major airports in the area. But no one had anticipated we'd be trying to set down our crippled airplane at a Chinese military airfield.

As we flew this slow circle, descending to about 2,000 feet above the sea, Regina would use her precision navigation equipment to give us an approach.

"Reggie," I asked her, "will you punch me direct?"

"You should see a city," she said cautiously, "then a river. The field is just to the left there."

Completing the 360-degree turn, I leaned forward and saw two green hills to the west and a curving beach northeast of the city. The scene looked like Kaneohe Bay on Oahu. The tan line of a concrete runway stood out beyond the first green hill.

"Lingshui," I called. "I have the field in sight."

As we crossed the white crescent of the beach at 1,000 feet, I tried to turn the wheel of the elevator trim tab, but it wouldn't move. A stab of hot adrenaline hit me. Something *was* damaged back on the tail. *Don't force it.*

Senior Mellos had been watching intently as Pat and I flew, scanning every instrument to make sure the good engines were performing as well as they could. All three of us on the flight station were absorbed with the landing we would soon have to attempt.

"I don't think we want to move these flaps," I said.

"Roger that," they replied almost in unison.

"Senior," I asked, "what are my speeds?"

As we approached the island, he had been methodically paging through the NATOPS, searching for solutions to our landing problems. But apparently the scenario that combined an unfeathered windmilling prop, nonfunctioning airspeed indicators, a missing nose cone, and a no-flap landing was not in the book. And none of us wanted to think about any damage that the nose wheel or its strut might have suffered when the Finback's forward fuselage slammed into our nose.

Crossing the palm groves and tan squares of the dry rice paddies at a thousand feet, the groundspeed registered 220 knots. We could probably trust that as a reasonable guestimate.

"One-forty-three knots is our no-flap landing speed," Senior said.

But since I had no way to judge our exact airspeed, we'd have to fly the final approach faster. "We'll add thirty knots to that," I said.

I hoped we really would not be setting this damaged plane down at an airspeed of 173 knots, but had the queasy feeling we were.

We still had almost fifteen tons of jet fuel aboard. If the nose wheel collapsed on landing, the fiery crash would probably be more catastrophic than swerving off a runway with blown tires on takeoff.

That's the way things would have to be. I bumped the power levers forward an inch, and our groundspeed increased.

The shimmering runway stretched out ahead, perpendicular to our flight path. I wanted to cross the airfield at this angle, flying at about 800 feet, so I could inspect the field. But after I crossed the runway, I would have to bank 30 degrees left into the dead engine to avoid the steep green slope of a hill. Back at Whidbey Island, flying an undamaged airplane with four good engines, I wouldn't have thought twice about the maneuver. Today, struggling with the buffeting control yoke to reach, but not exceed, the proper angle of bank, left me sweating.

We crossed the airfield at an altitude of about 700 feet. There were no airplanes or vehicles on the runway. I caught a glimpse of open aircraft revetments to the right, each enclosing an F-8 II Finback. But this was not a sight-seeing flight.

I held my breath as I banked left to avoid the green slope.

"Gear down," I told Pat, and automatically, "landing checklist."

Pat's hands stabbed down on either side of his seat. "Where's the checklist?" he asked frantically.

The checklist had been snatched up and dumped with the classified material in the flight station.

"Pat," I said, "we've got five set in the backend. Flaps are going to stay up. Get the gear down!"

Pat reached down to grab the tire-shaped knob of the gear handle and pulled it down. All three of us watched, holding our breath as the landing gear position indicator on the right panel clicked from Up to Down. The familiar rumble was much louder with the nose cone missing. The gear did not seem to have been damaged.

For the first time since the collision, the tight coil of fear inside my chest relaxed. *We're going to live*, I finally realized.

"And now we've got three down and locked," I called out.

"Yeah, right," he said, the relief clear in his voice even through the roar of the air stream, "okay."

I rolled out of the turn at about 500 feet and headed toward north to turn again onto final approach. I had no idea what the winds were, since we hadn't been able to speak to the tower. There had been no marker boards indicating the runway's distance, but it looked long enough to handle us since it was an F-8 base. It would have to be because we *were* landing.

Gripping the shuddering yoke only as tightly as I had to, I rolled back left into the dead engine to line up with the runway.

"Keep calling out my groundspeed to me until we're on the runway," I said to Pat. I wanted to keep well pitched down for a clear view of the runway. Once I reached the threshold at the end, I'd level off and chop the power to let us settle. But I wouldn't fly a standard nose-high flare, touching the main gear down first for aerodynamic braking before dropping the nose gear onto the pavement. With our damage, I wasn't willing to slow down any more and discover the low-spewed flying characteristics of this battered aircraft.

I was going to fly the airplane into the ground today, much as a Tomcat pilot hit the deck of a carrier. But at least I didn't have to try to snag an arresting wire. And there was another difference. If I did not stick the landing, I would not have another chance to fly a touch-and-go. The drag from the number one prop and the severed nose, combined with our weight, made that impossible. I had to get this airplane down safely on the runway the first time around.

"One seventy-five," Pat intoned, "one seventy . . . one sixty-five." The runway rushed toward us. I kept both hands on the chattering yoke, but was ready to chop power with my right hand.

"Whoa," Pat said. "Whoa."

We were almost down.

"One seventy," Pat announced.

The wide concrete blocks of the runway were beneath us. I eased back on the yoke and pulled the three power levers back to Flight Idle. The big airplane settled so gently, we could hardly feel the wheels touch.

Now we were screaming down the runway, and I had about fifteen seconds to slow the heavy aircraft. But I could not reverse the pitch of the three good props above 130 knots and I couldn't accurately judge our speed. So I had to wait for the green beta lights to blink on to show the prop pitch of the spooled-down engines had safely reached less than 10 degrees.

The second I had the lights, I gripped the three good power levers and pulled them into reverse, pumping the rudder to keep us on a reasonably straight path. The plane felt squirrelly, but we were slowing fast. "I've got it," I said, as much to reassure myself as the others. "I've got it."

When I felt the rudder going dead, I continued holding centerline by fanning the power levers and then took over control with nosewheel steering. I was acutely conscious of the near silence in the flight station, even though 50 knots of wind was still blasting through the ripped pressure bulkhead.

Suddenly I could hear the crew in the backend screaming and cheering with joy. I released a deep sigh. *I cannot believe we're alive.* I turned to silently share the moment with Senior and Pat. *I just can't believe we're alive.*

And then another, colder thought struck me. *We are alive, but we're also in China.*

## Chapter Five

# LINGSHUI AIR BASE, HAINAN

### 1 April 2001, 1038 Hours

I tapped the brakes lightly with the toes of my flight boots, slowing the plane to a gradual roll. The runway was a little bumpy in places, with grass sprouting between the cracks of the square concrete blocks. Off to the left, two open green military trucks sped past a control tower black with mildew stains. The trucks were full of soldiers.

"Looks like they were expecting us, after all," I said to Pat and Nick Mellos.

"I guess they heard the Maydays," Pat said.

It was obvious the Chinese had known we were coming. In fact, before Jeff Vignery had turned over the left seat to me following our recovery from the near-inverted dive, he had seen the surviving Finback fly parallel to us for a few minutes. Then the fighter had streaked away northwest toward Hainan. That also explained our reception committee.

As we neared the end of the runway, a skinny little guy in a short-sleeved shirt, his feet in sandals, stepped onto the concrete to wave taxi directions at us. He was walking backward, pumping his right arm frantically, indicating we should turn left onto the ramp where the trucks had now pulled up and the troops were jumping down. About a third of the soldiers carried AK-47 assault rifles. I followed the man's orders.

Even though the engines still roared through the shattered pressure bulkhead, it was much easier to hear over the intercom without the high-speed airstream shrieking past our faces.

"Ballgame," I asked Johnny Comerford in the back end, "have you called to report that we're safe on the deck?"

"Not yet," he replied immediately. "Give me a minute, okay?"

The secure radio link to our chain of command would have been "zeroized" during execution of the Emergency Destruction plan. It would take those back there a few minutes to briefly activate that piece of equipment.

I completed the turn, and our taxi guide thrust his palms out for us to stop beside the trucks. I nodded vigorously to demonstrate I understood and that I was cooperating, and then set the parking brake. Now the guy was slashing and chopping his hands, signaling us to shut down engines. The troops began to form a circle around the plane. Past the taxi ramp, there were a few more mildewed concrete buildings with orange tile roofs and stands of palms, their fronds moving in the light breeze. I saw more rice paddies and thatched houses on the surrounding hillsides, but this was definitely a military airfield. There was nobody out here except those armed troops and us. And they were used to having orders followed.

Johnny realized that time was critical. But I had to stall as long as I safely could without provoking any of those soldiers.

"Lingshui, Lingshui," I radioed on the distress frequency. "This is Kilo Romeo 919. We need a few minutes to let our engines cool."

That wasn't true, but I hoped it sounded convincing. I definitely *did* need time to adjust my mind from struggling for survival in the air to an altogether different kind of struggle on the ground.

*Keep stalling*, I told myself. We had to report what had happened so that our command would receive word directly from us, not a distorted version from the Chinese.

Now our taxi man was becoming adamant, pointing at each engine and slicing his hand across his throat. I could no longer risk any of the kids down there getting reckless with those AK-47s the way the Finback pilot had flown his fighter.

"Senior," I said, "shut down number four."

Senior Mellos pulled the power lever of the outer right engine back to Ground Start and chopped the fuel to the engine.

The guy on the ground seemed relieved as the number four prop stopped spinning.

That would satisfy the Chinese only a minute or two, I knew. But I didn't think they would try to put ladders up on the wings with numbers two and three still turning. If they did, the prop wash would blow them off.

"Johnny," I called again, "you get that call off yet?"

"Yeah," he answered, sounding frustrated. "But they want us to hold on a few minutes."

"Senior, shut down three and two," I said. "But keep the APU running."

I could picture what was happening in the chain of command. Our report was ringing alarm bells around the world, and the duty operator at the other end of the secure radio circuit must have been ordered to keep that channel open as long as possible. Unfortunately, our radio needed power, and that came from either the engines or the Auxiliary Power Unit located under the cockpit.

I could probably delay the Chinese a couple of minutes longer by shutting down the last two engines. But we couldn't risk stalling much longer. By now Johnny would have made certain that a clear, concise report had reached the U.S. Military: A Chinese F-8 Finback II had collided with us in international airspace. Our aircraft had been severely damaged. All the crew was safe. The Emergency Destruction plan checklist had been completed. And we'd had to land at the nearest airfield, Lingshui.

A high-wing blue-and-gray camouflage twin-prop aircraft taxied onto the far end of the runway, where it sat, rippling in the heat mirage off the concrete, while the crew ran up the engines. Then the plane rolled toward us for a smooth takeoff and banked away to the southeast.

"They're going to look for that Finback pilot," I said. I hoped they found him. He had almost killed us all with his reckless and sloppy flying, but after our own survival, I didn't want that pilot to die.

With the engines silent, the APU's little turbine seemed especially loud. Our excited taxi-ramp man was chopping the air again, signaling us to kill the APU.

"Go ahead and shut down the APU, Senior," I said.

I shucked off my seat harness and moved toward the rear of the plane. The deck was littered with smashed glass and torn wiring from the equipment that had been destroyed. Without the APU, the air-conditioning had shut down and the cabin was getting hot. Already the collar of my flight suit was soaked, so I pulled off my parachute and survival vest and dumped them in a sweaty pile on one of the torn-up in-line position stations. I wanted to be the first person to speak to the Chinese, who were now clustered beneath the main door.

But when I reached the door, a crew member was hunched over, vomiting into a paper bag.

"Do that back aft," I said. "I don't want to have to smell puke while I'm trying to talk to the Chinese."

Other crew members were still moving back and forth among the in-line positions, double-checking that the Emergency Destruction plan had been completed. I unlatched the door but did not unfold and drop its attached sectional ladder. A blast of humid heat swept over my face.

Below, a clump of men in casual uniforms stared up at me. One was an interpreter who spoke halting English.

"Do not move inside airplane," he said, his eyes shifting to the crew members who scurried past the door behind me. "Does anyone need medical assistance?"

I shook my head. "No, everyone is all right."

People still darted among the in-line stations.

"Do *not* move," the interpreter insisted, relaying the orders of an officer behind him. Another crew member ran past the door. "No one to move inside airplane," the man called up, his voice harsh now.

The officer spoke in Chinese, and the interpreter called, "Do not touch anything." He was sweating, obviously very nervous.

Again the officer spoke and the translation followed. "Do you have weapons? Give us your weapons."

I shook my head emphatically. "We have no weapons."

"You sure?"

"Yes."

The sun was beating on my head and I felt jangly from all the adrenaline that had pumped through my system. But we were in communist China, I was the mission commander, and I had to stay on top of things.

The officer was speaking urgently into a cellular phone. Another crew member passed the open door. "Do not *move*," the man shouted. The young soldiers with the AK-47s looked scared and alarmed.

I turned and called into the cabin, "Stop going past the door."

"You must get off airplane," the officer ordered.

"No," I said, shaking my head, "we do not want to get off the airplane. Can we please use your telephone?"

The officer looked like he had about had it with my stubbornness and spoke an angry burst of Chinese that the interpreter did not try to translate.

"Well," I said, "if I go with you, can I use your phone to make a call? I have to tell my command we are safe."

"That is not possible," the officer replied. "We will take care of that. Do not worry." The interpreter now beckoned for me to leave the airplane.

I could hear harsh orders and saw several more armed soldiers forming into a cordon between the tail and the wing. We had completed the Emergency Destruction plan checklist, and it was time to leave. The people on the tarmac were armed; we were not. If they wanted to beat us, or subject us to other forms of physical or mental torture, it was in their power to do so. They could starve us to death, or even shoot us if they wanted to. I did not want to begin our involuntary stay in China on any worse grounds than necessary.

"Hey, Ballgame," I told Johnny Comerford, who stood just forward of the door. "It's time to get off."

"Okay," he answered, "you're right." He cranked down the door's folded sectional ladder, and I was the first to descend.

One by one, the crew followed. I counted each of them as they came down, blinking in the muggy sunlight. Some were clearly intimidated by the nearby armed soldiers. Others kept their faces calm and neutral. Senior Nick Mellos was the last crew member off the plane.

The officer approached the folding ladder.

"You are not allowed aboard the aircraft," I said. "It's American property."

The interpreter nodded. "It is okay. We will guard it for you."

*Right*, I thought, *like the fox guards the henhouse.* "Can I lock it up?"

"Not possible."

While I'd been talking to the officer, a medium-size tan bus had pulled up to the base of the ladder. We were ordered to board the bus, and the interpreter directed soldiers to distribute two cartons of bottled water and packs of Bao Dao Chinese cigarettes among our crew. About half of them were heavy smokers, and they immediately lit up, filling the stifling bus with thick smoke. I could have ordered them to stop, but I knew how they must have felt, so I kept quiet and put up with it.

"Where are we going?" I insisted.

"Wait and rest."

We sat, swigging lukewarm water from large plastic bottles, breathing harsh tobacco fumes. The Chinese brought a small van, big enough to carry four of us at a time to use the toilet. I made sure an officer accompanied each group. If the toilets were any indication of the facilities here, our visit to Lingshui was not going to be very comfortable. The rough concrete urinal had backed up and the floor was a stinking flood. The doorless stall, a hole in the concrete with raised footsteps, was just as filthy. The air swarmed with flies.

Waiting on the taxi ramp for the Chinese to decide on the next step, Pat Honeck, Johnny Comerford, and I conferred at the front of the bus.

"What do you think's going to happen?" I asked them.

"I have no idea," Johnny said.

Pat looked grim. "Neither do I."

Some farmers in conical straw hats had approached the airfield from the nearby hill and were gawking at our big, mutilated airplane. But the soldiers trotted over, shouting angrily, and scared the peasants back out of sight. Again I had the sense that we were completely isolated here.

"Everything did go all right in the backend, right?" I asked Johnny again softly to reassure myself.

"Everything's good back there."

"Okay," I said. "We're still on the mission. Just so we don't confuse the crew and also the Chinese, as far as I'm concerned, I'm still the mission commander. That could change if things get physical with the Chinese and they start treating us like POWs."

"Okay," Pat said.

"That's fine with me," Johnny added.

I felt it was necessary to clarify this point because, under the Defense Department's Code of Conduct, the senior ranking officer assumed command of all Prisoners of War held at a particular site. And Pat Honeck, while also a lieutenant, had more time in grade than I did, which made him senior to me.

But I didn't consider ourselves to be POWs yet. The collision with the Finback had been an accident caused by poor airmanship and stupidly aggressive flying, not an act of war.

I stood up and turned to face the crew. Many faces were flushed with heat and excitement, some pale with anxiety. I knew what I had to say, and could only hope that I'd sound reassuring.

"Hey, everybody," I called above the buzz of voices. "I know this wasn't the ideal place to land, but it was the only choice we had. You did a great job and I'm proud of all of you. We're lucky to be alive. Remember that. We'll get through this together, just like we did on the plane. Keep the faith."

"You guys in the flight station did an incredible job," a voice called from the hot, crowded bus.

"Thanks for saving our lives," another added.

Senior Mellos leaned over and gripped my shoulder. "No regrets," he said. "I wouldn't have done anything any different."

As we waited for the next development, I wondered what the Chinese military and civilian authorities had in mind. At this stage, I could not imagine that they would find a pretext to detain us very long.

But I had been in China for only an hour.

Sitting in the humid heat, I thought about my priorities as a leader. First, I had to guarantee the crew's health and physical safety. Although some of the Chinese soldiers had been armed, no one had menaced us so far. In fact, the officer had seemed quite concerned about our comfort.

"Keep drinking water like crazy," I called back to the crew members. I didn't want anyone getting dehydrated, especially after the terror they had undergone aboard the crippled plane.

I wished I had more details about the message to the chain of command that we had sent over the secure radio link before shutting down engines and the APU. But I didn't want to quiz Johnny here on the bus. You never knew if the driver spoke English, and there might be bugs installed. I really did have to radically readjust my mental process now. We were no longer nursing a disabled airplane to a landing, but faced a more subtle kind of challenge.

I gazed out the window. There was no more breeze to move the palm fronds. Tall cane grass swept away outside the rusty fence toward the hills. A few more peasants had crept down through the grass to stare at us, but again the troops were hollering at them to disperse. In front of us I could see sloped tails of Finback fighters protruding from their humped earthen revetments. I knew this was the home base of the pair of fighters that had intercepted us. And now their lead pilot was down at sea. I wondered how good Chinese water survival training was. Did their aircrew carry miniature radio beacons like we did to help the search-and-rescue aircraft hone in on the signal? Was he injured in the collision? We would probably find out soon.

The interpreter climbed aboard the bus. "We will take you to eat now."

The mess hall was another low, mildewed concrete building. But its interior was actually air conditioned. We filed in the door a little apprehensively and were startled to see the Star Sports International broadcast of a soccer match on a big-projection TV screen. A kitchen staff in stained white jackets stared at us with their mouths half open. I realized we were probably the only Westerners they had ever seen.

Each of us was given a tin mess kit and the choice of either a fork or a pair of chopsticks. The lunch was pretty basic: several scoops of boiled rice, some stewed green vegetable, and hacked-up chunks of medium-size boiled fish. If you were lucky, you got the center section, if not, the head or the tail. I waited until everyone passed through the chow line before proceeding, because I didn't want to receive a nice piece of fish that someone else might have missed. Some people who got fish heads had been unable to eat them.

But Senior Mellos feasted on the leftovers. Old sailor that he was, he knew better than to turn down a square meal. "Never know where the next one's coming from," he muttered, spitting a piece of fish skull into his mess kit.

I managed to eat all my food, a little surprised at my hunger. Then the officer and the interpreter called me out of the mess hall.

"Write names of all people on airplane," the interpreter said, handing me a sheet of paper and a blunt ballpoint. The only place to write was a tree stump, so they let me back inside to use a table. Some of the crew already had their heads down and were sleeping, completely spent by the ordeal we had been through. I wrote slowly, also fighting waves of exhaustion.

While I wrote, two junior officers entered and introduced themselves as our new, "official" interpreters.

"Call me Lieutenant Tony," one young man said with a slight smile.

"I am Lieutenant Gump," the other announced.

*You've got to be kidding*, I thought. But that was no doubt the poor guy's name, since he had probably never seen *Forrest Gump*.

"We are going to take you for sleeping," Tony said.

We filed out of the mess hall, and again I left last, counting everyone as they passed me. The officers' barracks was a short walk away, a two-floor concrete building with flaking whitewash partly masking the pervasive black sheets of mildew. We were led up a flight of stairs to a corridor that stretched to the right along the second floor.

"You take rooms," Lieutenant Gump said, pointing.

I waited until people had picked rooms, and then took Pat Honeck and Johnny Comerford aside to share one with them. The rooms were all about the same, gritty floors, with chugging, ineffective wall air conditioners and gray sheets on the bed. In fact, there was hair on most of the bedding, an indication it had not been recently washed. But things could have been a lot worse. The crew was together and we could circulate freely along the corridor.

Once everyone had chosen rooms, I assembled the crew on an outside porch that was unlikely to be bugged and where I thought we could speak openly.

"Just watch what you say inside," I told them, pointing to my ear. "People are probably listening. Keep your cool and we'll get out of this together."

I turned to Senior Mellos. "Senior, let's keep a watch running tonight to make sure everything's going well."

"Roger that, sir," he replied, then turned to cryptographic technician Shawn Coursen, the crew's leading petty officer, the equivalent of an Army platoon sergeant. Setting a watch was standard Navy practice to keep track of people in a crisis.

"Petty Officer Coursen," Senior said, "I want you to set up a watch tonight."

I knew I didn't have to say any more. Senior Mellos and Petty Officer Coursen would handle the details. Within half an hour, Coursen had drawn up a roster to keep two people awake each hour on the opposite sides of the corridor. They would be able to monitor the crew's movements as they went from their rooms to the toilets at the end of the corridors. These were similar to the reeking, filthy stand-up stalls we'd seen on the airfield.

Later, a doctor arrived to look at the crew members who had complained of headaches, which had probably been triggered by the explosive decompression.

Senior Mellos, the other pilots, and I remained on the porch for a private discussion about what to expect next.

"Well," Senior said, puffing on one of the Chinese cigarettes, "we're going to be interrogated."

"We've all been trained for that," I said, thinking about my own training in the frozen Maine woods. "We know what we can and can't say."

Everybody nodded grimly. The Chinese had a reputation for brutal interrogation of their own people, but we could not predict how they would treat us.

"Look," I said, "we had a crash out there and they're going to be very curious. You don't want to volunteer too much, but they need to know the truth. We have to give them that."

Once more, the four men around me accepted this logic.

"So don't be surprised if they want to start with the flight station," I added. "Let's hope we can keep it limited to that."

But if I could not prevent the Chinese from interrogating the backenders as I intended, I wasn't too worried. The enlisted people flying on EP-3Es were among the Navy's elite. Most either already had a college degree or would soon earn one. They did not need guidance from me on the details of what they could and could not say to the Chinese. Instead, these young sailors looked to me for leadership.

The evening meal was similar to lunch, with the spinachlike vegetable ladled atop the rice and a watery soup instead of fish. But there were also mangoes and sweet little bananas. Apparently, we were being fed after the Chinese personnel, because we had the mess hall to ourselves. I had to encourage people to finish their meals. We might be here for several days, and I didn't want anybody becoming weak and vulnerable.

After the muggy rooms, the air-conditioned mess hall felt good. But then it was back to the barracks. At around 2000 hours, Lieutenant Gump came to the room and announced, "Now to go sleep."

But that was not so easy. The air conditioner clanked, but spewed out only hot air. And mosquitoes rose from a stagnant pool outside to whine through inch-wide gaps in the wall between the air conditioner and its frame. The situation reminded me of the "luxury" rooms we'd had in Ecuador, where the air conditioners had blown out hot air, and we'd had to stuff the gaps with wads of toilet paper to hold back the hungry mosquitoes.

As exhausted as I was, I just could not sleep. My head on the sweaty pillow, I fought waves of troubled thoughts. The frightening images of the collision and near-fatal dive kept intruding, and I'd sit upright. But when I did manage to push those pictures aside, I couldn't form a logical projection of what was going to happen to us next.

I was still muddling through all this when the door opened and Lieutenant Tony appeared to shake me by the shoulder. "Lieutenant Osborn, we must talk to you."

I wasn't wearing a watch, but I saw by the light from the corridor that it was exactly midnight on Pat's watch. "Where?" I had tried to keep my voice neutral, but I know my anxiety slipped through. *Where are they taking me?*

"Right here on this floor."

I shook Pat awake. "Meeso, I'm going for interrogation. Pass the word."

As soon as I left, Pat logged the time and told the crew members on watch.

• • •

Lieutenant Tony led me down the corridor, past the stairs, and to the door of a larger, brightly lit room. But there he stopped. He looked uneasy, but nodded for me to enter, then closed the door. It was hot inside, and smoky with the same biting stench of Chinese cigarettes. The glare dazzled my eyes, and I realized I was looking at the lights of three separate video cameras set up on tripods to focus on a single straight-back chair before a table. A young man with a large 35 mm still camera stood patiently beside one of the video operators. The Chinese had apparently prepared for a rich propaganda session.

"Sit," someone ordered.

Two interpreters sat beside two senior officers behind the table. One was introduced simply as "base commander." The other was the officer who had commanded the troops at the plane. I had a mental picture of him as the equivalent of a U.S. Navy lieutenant commander, even though he wore a plain short-sleeve shirt with no rank.

Squinting into the camera floodlights, I realized I had been awake for almost twenty-four hours. And from the way this room had been set up, I knew this interrogation would not be brief. They had intentionally chosen the middle of the night, expecting I'd be at my weakest.

The base commander began. "We are here to investigate your ramming of our aircraft. We lost the pilot and have yet to find him."

The interpreter's thick accent was difficult to follow.

The younger officer beside the base commander glared at me.

"We need to find details of your mission for investigation," the base commander continued. "Are you willing to give us names and positions of every people on board?"

The second interpreter translated this, adding choppy language to the accent problem.

*This is not a straight accident investigation,* I immediately realized. "I'm willing to give the names, ranks, and Social Security numbers of every person on the airplane," I said, speaking clearly in a neutral tone. "But I have to talk to the American ambassador in Beijing and to my chain of command."

"We need positions of every people on airplane," the younger officer insisted.

Keeping my voice even, I tried to deflect his demand. "Once I tell my superior officers we are all safe and receive guidance from them, I will be able to cooperate with the investigation of this accident." I had emphasized the word "accident" so that I didn't seem to be casting blame for the collision on the Finback pilot.

"We need names and positions of your crew," the lieutenant commander persisted, speaking through the first interpreter.

"I will give you the names and positions of the crew who were flying."

The interpreter handed me another piece of paper and a scratchy ballpoint. I wrote the names of the pilots and flight engineers. If this was a legitimate accident investigation, the Chinese had a right to speak to the flight station crew, because we could each provide needed details of the incident.

But I also knew they had their own eyewitness, the surviving Finback pilot, who had been out there on the left, trailing our airplane, and would have seen the whole collision unfold.

"Why you ram our plane?" the base commander now demanded. "We have witnesses who saw this happen."

So much for the second Finback pilot providing an honest account. But I understood the extreme pressure he had faced to give a favorable account of the collision to his superiors. And even if he had told the truth, there would be no way they would accept fault.

I was painfully aware of the glaring video lamps and the swirling cigarette smoke. But I kept my voice calm and tried to speak in clear, even sentences. "We did not ram your aircraft. We were southeast of Hainan Island, operating well out in international airspace and heading toward the southwest. That was when I saw the two fighters. One approached very close to our right wing, but when we turned left toward home and a new course of 070, they came in from the left. One flew with a very fast . . . at a high closure rate . . . and attempted to arrest that rate of closure." I lifted my hand, using a typical airman's descriptive gesture. "He raised his nose pitch." My fingers shot up. "That's when he impacted my number one propeller."

I had spoken slowly and calmly, knowing that I had to avoid being confrontational or fault the Finback pilot for his reckless and unskilled airmanship, as this would only provoke a more aggressive approach to the interrogation on their part.

"No, you turned left into him," the younger officer said.

"That is not correct," I replied. "We were flying on autopilot. My airplane never turned until after the collision."

Now the base commander gestured. "No. If our pilot hit your left wing, why your airplane not turn right?"

*Either these people are amateur aviators,* I thought, *or they're trying to make me admit I turned left before the impact.* "My nose pitched up when your fighter struck the propeller," I said, raising my hand to indicate the snap roll. "That impact was what caused the sudden roll to the left."

"You turned left to hit our airplane," the younger officer repeated. Obviously, the Chinese did not want to understand the aerodynamics involved.

"No, after the impact."

"We have several witnesses who saw you turn left," the officer said coldly.

*Several?* Then I realized he must mean the ground radar operators tracking the intercept who saw the blips converge and my plane roll suddenly left. I shook my head. "We did not roll left until after the fighter hit the propeller."

"Liar!" the base commander bellowed, slapping the table. "You had better tell the truth . . . for the safety of your family, and your crew . . . and yourself. You must cooperate with this investigation, Lieutenant Osborn."

Despite the smoky heat, I felt a chill. And tired as I was, his message was clear. The Chinese were threatening my family back in the States, my crew here on Hainan, and myself, with physical harm if I did not "cooperate." But they did not want the truth; they wanted correct answers.

"Tell why you rammed our airplane," the base commander repeated.

I was annoyed, but also nervous. I had not been deceptive—although, to avoid confrontation, I had not mentioned the Finback's first two passes before the collision—and I intended to keep simply telling a truthful account of the incident, even if it did not meet their idea of a perfect accident investigation. However, if they thought threatening my family or my crew would help their "investigation" or scare me into admitting something that was false, they were dead wrong.

Slowly, I reiterated the description of the Finback's last catastrophic pass, ending with the impact on our number one prop. "Then your aircraft broke in two pieces. My airplane pitched up, rolled hard left, and went into an uncontrolled dive."

"Why you turn left into our airplane?" the younger officer asked as if he had not heard the interpreter translate my words.

They continued to repeat their questions. And I continued with my original answers. Not having a watch, I had no idea how long this session had lasted. But I did know my flight suit had become clammy with sweat and my eyes burned from the video-light glare and cigarette smoke.

"You must cooperate with investigation," the base commander said yet again. "Remember about your safety."

I was just about exhausted, and a wave of fear swept over me. Were they going to take me away to a cell and start beating me? What if they isolated me from the crew and just kept up these sessions around the clock? Worse, the base commander had threatened the "safety of your crew." What if the Chinese began to torture crew members in front of me? Would I cave in and agree that we had caused the accident by turning into the fighter?

Through the glare of the lights, I tried to judge from the two officers' expressions what they themselves actually believed. Their faces were impossible to read. The surviving Finback pilot had either lied, or these officers would not accept a true account if he had given one. Maybe they themselves had been watching the intercept unfold on the ground radar screen and actually believed we had been at fault. But I did not think so. The answer they sought had little to do with the real event and a lot to do with internal Chinese and international politics.

If they could pressure me, the mission commander, or a member of my crew, into a videotaped confession that we had caused the collision, their "investigation" would have achieved a number of goals. The Chinese would have been absolved of reckless flying, and the blame for the incident would rest with the Americans. That would place the Chinese military and civilian leadership in a strong position with their American counterparts. I couldn't know everything that was happening between our two countries at that moment, but I did know the day's events had to have triggered a major crisis.

So I just would not cave in and give them what they wanted. Maybe if they did get physical and resorted to actual torture, I'd have to reconsider. But at this point they were going to have to literally beat any acceptance of blame for the accident out of me.

"If you want to go home," the base commander said, "you must cooperate with investigation."

"I have cooperated," I answered, my voice hoarse with fatigue.

"Why you ram our airplane?" the younger officer asked again.

*Your technique is getting old,* I thought, then launched into another detailed explanation of how the collision had evolved.

"We know you from Kadena," the base commander suddenly said, trying to throw me off.

"Yes, we're from Okinawa." This was no secret. The Chinese had been tracking our flights for years.

Now they shifted gears again. An interpreter handed me another sheet of paper and the pen. "Draw accident and your flying position from Okinawa." They wanted the track line. But I wasn't about to give them that. I drew a crude outline of Okinawa and another of Hainan. Then I sketched a dotted line south from Kadena to the approximate point of collision and drew a sloppy stick airplane to represent the EP-3E.

"We must speak entire crew," the younger officer said.

I wiped the sweat off my face. "That is not necessary because there are no windows back there where they were sitting and they would have nothing to add."

The base commander's face was passive, but I caught a different tone to his words before the interpreter translated. "You were on right. You not see everything."

"Correct," I conceded. "You can talk to the other two pilots and the flight engineers. They are the only ones who count. The rest were not at windows and cannot help you with your accident investigation." I took the chance of using their ploy that this interrogation was only about the collision.

"We need talk to every people to have information about why you flying there," the base commander replied.

"That is not relevant to the accident investigation," I said stubbornly. "I am willing to cooperate. But those other people have nothing to add."

The officers at the table gathered up the two pieces of paper I had written on and spoke between themselves in rapid Chinese.

"That is all we have for now," the base commander said.

This first interrogation lasted about five and a half hours. The sun was rising as I tiredly climbed the chipped concrete stairs to the second floor. That night Pat Honeck and Senior Mellos had also been taken for questioning.

We met back at the room at dawn to compare notes. Each of us had been subjected to the same repetitive line of questioning and blatant threats. But each man had kept his cool and truthfully answered every question about the intercept and the collision.

"It doesn't look like they're interested in getting to the truth," Senior said.

I nodded. This was a terrible embarrassment to the Chinese. Lingshui was the home base of the F-8 Finbacks responsible for the recklessly aggressive intercepts that had triggered the formal protest from Washington in the winter. But the Lingshui pilots had obviously ignored the American protest. Now one of them was missing, an American aircrew had narrowly escaped with their lives, and a crippled EP-3E sat on the Lingshui taxi ramp like a beached whale.

"Well," I said, "they have all the information they need for a real accident investigation, and we're not going to give them anything else. Right now I want to brief the crew about the interrogation. Then I want to get some sleep. I'm wiped out."

When my head finally hit the musty pillow around 0630, I felt like I was being pulled under a warm, dark river as sleep swept over me.

But I didn't sleep long. Jeff shook me awake only two hours later.

"Time to wash your face for breakfast." My flight suit smelled like a football uniform left unwashed in a locker over a summer practice weekend.

"We need some clean clothes," I told Lieutenant Tony.

"They are coming now."

But first we had to assemble for breakfast. Once more I stood at the bottom of the stairs, counting the crew members as they passed. I smiled and tried to cheer them up, but I know I looked like hell, a day-and-a-half stubble, black circles under my eyes, and hair matted with dried sweat. Still, things could have been a lot worse. About this time the previous morning, I didn't think any of us would live through the day.

The mood had definitely chilled since the overnight interrogations. Now we were escorted between the buildings by armed guards. But the air-conditioning in the mess hall felt great, and I was even hungry for the exotic breakfast of small fried fish and green vegetables on top of rice. We also had little cans of sweet coconut milk. Not an Egg McMuffin at the McDonald's near the main gate at Whidbey Island, but, after all, we were a long way from home.

The guards were watching us closely as we filled our mess kits and went to the tables.

"Sit," they commanded. "Eat. Eat."

But I had instructed the crew that no one was to sit until everyone had food. In a situation like this, little habits of discipline kept up morale. And Jeff Vignery had suggested leading us in prayer before meals. That was fine with me. We were going to continue needing all the help we could get.

"Lord," Jeff said as we all bowed our heads, "thank You for protecting us and for this food. And please Lord, protect and comfort our families. . . ."

"Sit. Sit," a guard commanded. "Eat. Eat." But then Lieutenant Tony explained that we were praying and the man fell silent in a mixture of embarrassment and wonder. He had probably never seen people pray like this before.

Back at the barracks, the guards laid out stacks of the civilian clothes we had requested. They turned out to be bright blue-and-white athletic shorts and tank tops emblazoned with the white Nike swish and a logo reading "Official Replica."

"I am ready to go back and face the interrogators," I said, laughing when I caught sight of Senior Mellos's hairy torso half covered by the skintight tank top.

We managed to coax a little cool air out of the wall unit, and I might have gotten another thirty minutes' sleep before I heard voices in the hall. The guards had come back to take me for more interrogation. But Johnny and Pat were out there, arguing with them.

"He's too tired now," Pat Honeck said.

"Lieutenant Osborn has told you everything you need to know," Johnny said, really sticking it to them. "He has to rest now." The Chinese reply was indistinct, but I heard footsteps moving away down the corridor.

The next thing I knew it was noon and Pat woke me from my restless doze. "They've taken Jeff and Regina for questioning."

I shook myself fully awake. This did not present a problem, because both Jeff and Reggie would only substantiate what we had already told the Chinese officers. But I did not want them subjected to prolonged interrogation. As it was, they returned just under two hours later and related an experience similar to that of Pat and Senior. If the Chinese now continued to doubt the valid account of the collision, that was their problem. We had all given them the truth.

Lunch that day was rice with a lot of vegetables, somewhat tasteless, but no doubt nourishing. Again we waited until everyone had a meal, because I wanted to be sure there was enough food for everybody. Once more we said a prayer before sitting to eat. Again the guards looked at us like we were crazy. But I think they had also begun to respect us for our faith and discipline.

Looking at the faces of the crew members seated around me, I saw that some were tight with concern and others flat-out scared. It was one thing to attend training classes about this kind of experience, something altogether different to live through it. *Boy,* I thought, *it could be a long while before we get out of here.*

But all of us took great comfort in the fact we had strong leadership backing us at home. As Johnny and I had explained to the crew, we had gotten word to our chain of command that we were safe on the ground at Lingshui following the collision. And we all trusted that chain of command, right up to the top. I knew with certainty that the Bush administration would not abandon us out here and had faith in the president to act decisively. Vice President Dick Cheney was a former secretary of defense who had overseen complex geopolitical and military negotiations during the Persian Gulf War. Secretary of State Colin Powell was a man of his word, a seasoned Army combat veteran who understood from personal experience the extreme hazards that members of the armed services regularly face. Donald Rumsfeld, the present secretary of defense, was himself a former Navy pilot. He would know what we were going through. And the president's national security adviser, Condoleezza Rice, was a brilliant academic, but also reputedly shrewd and tough. If any team could get us home, it was them.

If that team in the States weren't enough, the American ambassador in Beijing was retired Navy Admiral Joseph Prueher, the former commander of American forces in the Pacific. He understood the Chinese government and military well. And I knew his staff was working hard for us.

There could be no doubt that the American team was strong. But I also had to face the fact that the Chinese would not easily accept the truth about the collision. The goal of their investigation was to force the crew and me into admitting we had caused the accident. And they were no doubt willing to take as much time as necessary to meet the goal.

At 1730 hours, they came to take me for more interrogation.

"You must hurry," Lieutenant Tony said. "Our rear admiral has arrived to talk to you. And they are waiting."

The new, larger interrogation room was just as hot and smoky, and the lights were still uncomfortably bright. Again Lieutenant Tony was denied entrance. The admiral behind the table was a silent, older man dressed in a plain T-shirt and rumpled slacks. The base commander glared at me from one side of the admiral, the younger interrogating officer glared from the other side. The same two clumsy interpreters sat at the ends of the table.

"Why you ram our aircraft, Lieutenant Osborn?" the base commander began, starting the whole damned line of questioning all over again.

This was unacceptable, and I felt like jumping up to storm out of there. Then I thought of my mentor, Lieutenant Norm Maxim. He would keep his cool in a situation like this and assess his strengths and weaknesses before going off half-cocked.

The flight station and I had already given them a detailed account of the collision. Hell, we had even gone easy on the Finback pilot. I needed to understand the limits of these interrogations. How far were the Chinese prepared to go?

Suddenly, I visualized a scene in the movie *Good Will Hunting*, where the troubled kid, Will Hunting, played by Matt Damon, describes to psychotherapist Robin Williams what he had suffered from his abusive father. Every morning, young Will's dad would lay out on the kitchen table a switch, a belt, and a wrench, then force Will to choose what he wanted to be beaten with that day. Finally, Will had picked the wrench. "Why?" Robin Williams demanded, amazed. "Fuck him," Matt Damon's Will replied. "If he wanted to kill me, he was going to do it anyway." Or words to that effect. I couldn't remember a four-year-old American movie in this sweatbox of a room on Hainan.

But I did understand the relevance of the powerful dramatic scene to my current situation. If the Chinese wanted to get rough and use the equivalent of the belt or the wrench, I could not stop them. Well, fuck *them*. It was time to find out now if that was the way things were going to be.

I turned in my chair to face the base commander directly. "I already told you everything."

"We know what you say," he replied coldly. "We need more details."

I shook my head abruptly, and felt the tension in the room increase. Was this when they'd start to rough me up? Well, the wrench was on the table and I had just pointed to it. "You already have enough. These cameras were here last night. Watch the videotapes."

The base commander could not hold in his anger. "Tapes do not contain enough detail."

I looked around the room and pointed to one of the cameramen. "That same man was taping last night and I told you everything I know."

"For the safety of your family," the base commander screamed, "yourself, and of crew, you must cooperate with investigation and answer all questions."

For a short man, his voice was surprisingly loud. And the interpreter's unemotional broken English did not convey the anger of the outburst.

"You *have* all the details," I said stubbornly.

Again the base commander screamed. "We have lost pilot and not yet found him. You crashed into his airplane and we must know exactly what happened."

"You do have every detail," I insisted. "I need to talk to my chain of command."

The base commander was trembling with rage, which was either authentic or a damned convincing act to impress the admiral. "You are a horrible leader, Osborn," he said, "a horrible officer. Why endanger your crew? Why you engage in espionage against the Chinese people? You are master spy."

The room fell silent. And I noted that the base commander's fingers were shaking when he lit a fresh cigarette. He looked up at me through the cloud of smoke. "Lieutenant Osborn, you will not be allowed to see your crew or anyone else until you cooperate. You need time alone to think."

The guards took me to another room downstairs. If anything, it was even dirtier than the rooms on the second floor. But this air conditioner was apparently set at sub-zero. It seemed almost freezing in the room after the heat of the floodlights. Outside, it was dark and I had no idea of the time. I did know, however, I would take this opportunity to stockpile any sleep I could. That was not going to be easy. No sooner had I lain back on the bed than two young guards pulled up chairs beside me and began chattering back and forth and blowing clouds of foul cigarette smoke toward my head. I jerked the collar of my T-shirt over my face. But the neck of my flight suit was sopping with sweat and the icy blast from the air conditioner made me shiver. Somehow, I slipped into mindless sleep.

As Johnny Comerford later explained, the first of the crew's courageous near rebellions began at 1940 hours that night, when guards came to the second floor to assemble everyone to go to dinner. "Where is Lieutenant Osborn?" Pat Honeck asked.

"Is he finished with questioning yet?" Johnny added. They were standing in the doorway of the room, hesitant to assemble for dinner.

"Lieutenant Osborn is not going to eat with you," the guard officer said through Lieutenant Gump.

"What do you mean?" Pat asked. "Why isn't he going to eat with us?"

"We need to eat together as a crew," Johnny said emphatically, "everyone together."

"He has been taken away to another part of the base," the guard officer said. "Don't worry about him. We are taking care of Lieutenant Osborn. Everything is fine."

"That is *not* all right with us," Pat said.

"We need to eat with him," Johnny repeated. "It is very important to us."

As the argument grew more heated, other crew members stood in their doorways, listening.

Finally, Johnny announced, "We are not eating until we see him."

"That is our decision," Pat said. "We will not go to eat until we can at least see and talk to Lieutenant Osborn."

As the guard officer left angrily to report this unexpected development, Johnny called after him, "Lieutenant Osborn needs to come back and stay in this room with all of us."

Then he turned to Pat and spoke in a lower voice. "I wonder if they're just going to start picking people off by rank, starting at the top?"

They knew where the original interrogation room was and stood in the hall, listening intently for any sound of physical abuse coming from it. What was going on down there? Had they really taken me to another part of the base, or was that just a lie?

One thing Johnny now realized was that having the crew eat three meals a day seemed very important to the Chinese. They wanted everyone to appear healthy. That knowledge was a bargaining chip for the future. Besides, no one was going to starve tonight because, out of habit, several crew members had stashed fruit and cans of coconut milk from the mess hall, which could be shared.

"Either the guards come back with guns and force us to eat," Johnny told Pat, "or we have to keep standing our ground here."

"If they actually point guns at us," Pat agreed, "we'll have to give in, but not before. Let's assemble the crew for a meeting. I want to talk to them all."

As the situation required, Pat had now assumed his role as senior ranking officer.

Once they had jammed everyone in their room, Pat briefed the crew on the latest developments. Those who had food to share went to fetch it.

Then Jeff Vignery began to lead the group in prayer. While this prayer was under way, a younger, unfamiliar Chinese officer in civilian clothes, accompanied by a video cameraman, arrived. "We want you to come to eat," he demanded.

"No," Pat said defiantly, "not until we see Lieutenant Osborn." Lieutenant Gump seemed nervous translating these words.

The officer glowered at Pat, then pointed at Jeff Vignery. "We want to talk to the other pilot."

"The last time you took one of our people to be interrogated," Johnny said, "he did not come back, and still has not. We are not going to let you take another until we get Lieutenant Osborn back."

The new officer glared at them, then stalked off. Jeff continued his prayer and the meeting ended.

About an hour later, Lieutenant Gump brought a printed note from the base commander officially ordering them to send Jeff Vignery for questioning. Pat and Johnny again refused. They had just gotten back to sleep, when they were summoned to the interrogation room at the end of the hall. There the base commander and the rear admiral ordered the crew to go to dinner.

"We want to eat," Pat told them, "but we are not going to do so until we see Lieutenant Osborn."

"We were told he is all right," Johnny added, "but we *need* to see him."

The base commander replied coldly. "He has been put into a new room so that he can improve his attitude."

Johnny and Pat exchanged a tense glance. Even in Lieutenant Gump's soft voice, those words sounded ominous.

"We need to see him," Johnny insisted. "We must be sure that he has eaten and that he is in good health."

I guess this was their equivalent of pointing to the wrench on the table. If their defiance was going to result in physical retribution, it would probably come now.

The argument lasted about twenty minutes. Finally, the base commander gave in. "All right, just the two of you can see him."

They discovered that I was not in another part of the base at all, but directly below in Room 106.

I was finally sleeping like a dead man, having filled my stomach with a mess kit full of rice and vegetables, when someone was shaking my shoulders. I pulled the chilly T-shirt off my face and blinked into the glare of a video lamp. *What the hell's going on?* As I shook away sleep, I saw there were at least five people in the room, including Pat and Johnny.

"Hey, Shane," Johnny said, again squeezing my shoulder. "Wake up, man. Are you okay?"

My mouth tasted foul. "Yeah," I muttered. "I'm okay."

"Have you eaten anything?"

"They brought me some food in here a while back."

The two guards were still leaning close, blowing their goddamned cigarette smoke across the bed. I fought the urge to jump up and wring their necks. But anger would not help. A scene like that captured on videotape was all the Chinese needed. Again I was reminded of Norm Maxim: Always think of the big picture, Shane.

Pat and Johnny explained the hunger strike upstairs.

"We wouldn't eat because you had been separated and we demanded that you be returned," Johnny said.

"Sounds good," I said, still groggy. "I'm okay. They just kind of cordoned me off down here."

"Keep the faith," Pat said as they left.

"Take care of the crew and don't worry about me," I said. "I'll be just fine."

Back on the second floor, the base commander told Pat and Johnny, "Now we want you to eat your dinner." It was almost midnight.

"Most of our crew is trying to sleep," Johnny said. This wasn't true, but it made a convenient excuse.

"We're really very tired," Pat added, "and we just want to sleep."

"We will eat breakfast with Lieutenant Osborn," Johnny said. "But we don't want to eat dinner. Or any other meals without him," he added. As soon as the two senior Chinese officers had left, Pat and Johnny spread word to the rest of the crew that I was okay.

Back in my room, the guards drew their chairs even closer and began an endless singsong conversation, punctuated by giggles. Their voices and cigarette smoke were really getting to me. That was the idea, of course.

The Chinese were not going to smack me around with a wrench. They were just going to try to break my will through sleep deprivation.

Well, I thought, grimly recalling those endless months in high school after the car accident when I had plucked the inward growing eyelashes, they didn't know how stubborn I could be.

Chapter Six

# LINGSHUI AIR BASE

## 3 April 2001, 0830 Hours

The night had seemed endless. Whenever I'd fall asleep, the guards would scrape their chairs and blow smoke in my face. Finally, I just lay there with my stinking T-shirt pulled up to my forehead, waiting for daylight.

Around 0800, two guards and Lieutenant Tony arrived with my toothbrush, some soap, and a small towel. There was already a plastic washbasin in the room.

"You go now for your breakfast," Lieutenant Tony said.

"Where's the crew?" I asked. "I'm not going anywhere without them."

During the angry, sleepless night, I had decided on a firm principle: I would do everything I could to stay in contact with the crew. If the Chinese could succeed in separating us physically, they would try hard to break our cohesiveness.

"I eat with the crew," I said, turning my back on Lieutenant Tony.

He left to report this latest outburst of intransigence.

That morning, when the crew assembled in the corridor upstairs to go to breakfast, Johnny Comerford went straight to the guard officer to demand that I be allowed to eat with them.

"Not possible."

"Well," Johnny said with a straight face, "the base commander told us last night that we could see him a few times a day."

"No," the officer said. "It was agreed that you could see him once every two days."

Pat shook his head. "No, it's a few times *every* day."

Another haggling argument ensued, with the Chinese apparently confident that the Americans would be hungry enough to cave in quickly. The Chinese were wrong. Pat and Johnny held their ground.

Once more, the two sides reached a compromise.

"All right," the guard officer said. "Lieutenant Osborn can eat with you, but at a separate table. He must be alone."

Pat and Johnny led the crew down the stairs for breakfast.

I met them at the foot of the staircase and once more counted as they filed past.

Johnny hung back and whispered. "How's it going? We were really worried last night that they might have drugged you."

"I'm okay," I whispered back, swallowing a yawn. "They just kind of kept me awake all night, but I did get dinner."

"Well," Johnny said, "that's more than we got."

As we moved along at the back of the line toward the mess hall, Johnny briefly recounted the crew's defiance. "And we're not supposed to talk to you now," he said as we reached the door.

Inside, it was obvious the guards were confused and nervous at the crew's rebellious mood.

Once we had our rice and fish, Johnny nodded to the chair beside his. "Just stand here next to me and maybe they won't say anything," he said, speaking from the side of his mouth like a convict in an old-time prison movie.

"Sit," the guard officer shouted. "Eat. Eat."

Today Pat led the prayer.

"No talking," the officer yelled.

We ignored him. Only when the prayer had ended did we sit. And the crew also continued to speak quietly as they hunched over their mess kits. After a few minutes, the frustrated guard officer stalked away in angry frustration. As soon as he left, the Chinese ordered us to hurry though breakfast.

Back at the barracks, I was taken to the larger interrogation room, where the rear admiral and the base commander were waiting, both puffing on cigarettes. There was only one video camera.

"We need more details of collision," the heavily accented interpreter translated for the base commander. "You have not told everything."

"I have nothing more to say." My voice was hoarse from the chill blast of the air conditioner and breathing tobacco smoke all night. I felt light-headed.

"We must complete a thorough investigation," the base commander said. "You have not cooperated."

"I *have* cooperated," I answered, struggling to control my anger. "I have *no more* details."

"You must cooperate with the investigation if you want to go home," the admiral said.

"We have all cooperated."

"For the safety of your crew, of your family . . ." the base commander began reciting the same old crap.

"I cannot cooperate any more than I already have."

Time lost meaning. There was just the hot, smoky room and the glare of the single video camera lamp. But I know several hours passed in this way. The only good thing I could see coming from it was that as long as they had me in the hot seat, the two principal interrogators could not hammer on another crew member.

Finally, the base commander shifted tacks. "Would you like the opportunity to see your airplane?"

"Sure," I said, a little wary.

"What caused damage inside airplane?" the rear admiral asked through the brighter interpreter.

Obviously, they had been all over the backend and seen the havoc caused by the Emergency Destruction plan. "No one is allowed inside," I said to deflect the question.

"What caused damage inside?" the admiral repeated.

"I was in the flight station the entire time," I said. "I have no idea what caused that damage."

"We will inspect the airplane together," the base commander said.

I remained in my chair. "Before I agree to go with you to the aircraft, I must insist that there be no non-American aircrew inside." I could picture the propaganda video crews taping the "master spy" on a choreographed tour of the plane's interior.

"We have right to enter," the base commander insisted. "The airplane is on Chinese sovereign territory."

*Be flexible,* Norm Maxim had always taught me. *Make the best of the situation.* What the hell? The Chinese had already crawled all over the inside of the plane. But I again tried to retain some control of the situation. "Well, that airplane is sovereign American property and you have no right to enter it."

"We have the right to enter," the base commander said. "And you must also enter."

They obviously wanted me to go with them. So maybe I could use this to extract a concession. "I will go with you to the airplane only if I can see the American ambassador to China or someone in my chain of command."

The base commander rose and smiled for the first time. "We agree. If you go to airplane, I give you my word you can see American representatives by midnight tonight."

This was a good compromise that gave me the opportunity to assess the extent of damage to the aircraft before meeting face-to-face with U.S. representatives. I realized, of course, that the whole session on the taxi ramp would probably be a propaganda event. That explained the absence of two of the video cameras. But getting a firm commitment from the Chinese to meet with American officials was worth any propaganda advantage the Chinese might gain.

Before leaving with the admiral in his Toyota SUV, I requested that I speak with the crew. I found them in Pat's room in the middle of an "all hands" meeting.

"They want me to see the aircraft," I told the crew. "I will not allow any non-American inside. But they would like Lieutenant Kauffman and Senior Mellos to come with me."

The Chinese had requested Regina come because they wanted to see her charts. I did not tell them the charts had been shredded and dumped.

On the way to the airplane, I sat in the backseat with the admiral. He surprised me by turning in the seat and speaking in better English than the interrogation interpreters. "So, you say the fighter came into your airplane like this and hit you?" he asked, making the same hand gestures that I had used to describe the collision.

He was depicting the accident exactly the way it happened. I matched his hand gesture. "That was how the fighter hit us."

The admiral nodded and looked away. He *knows* what happened, I thought. It's just a matter of politics from now on. I was sure the condition of the airplane would also provide physical evidence of how the collision had happened. But there was not going to be any independent investigation to weigh that evidence.

Instead, we found a miniature Chinese media dog-and-pony show set up in the brutal sun in front of the airplane. The two interrogation video cameras were there, as well as several television reporters and their crews. The admiral assembled everyone at the plane's shattered nose and began speaking in rapid Chinese while the reporters scribbled notes.

I was surprised when the interpreter translated the admiral's words. "This is the pilot of the American aircraft that smashed into our fighter and sent it crashing into the sea." The interpreter's tone was neutral, but the admiral sounded outraged. "He then violated sovereign Chinese territory by landing without permission."

That was the Party line, all over again.

Senior Mellos immediately turned his back and walked away. After the admiral had made a statement and answered several questions, he turned to me and said through his interpreter, "Now we will see inside of airplane."

"I won't be able to help you," I told him. "And you won't be able to go on the aircraft."

The admiral scowled. Reggie and I also turned our backs and walked off toward the number one engine. I didn't want to continue playing the "master spy" for the Chinese media. The admiral could yak away all he wanted about espionage. The hell with that. We had given them some video footage in exchange for the promise to meet with American representatives by midnight that night.

The admiral seemed pressed for time, and when he realized we would assist no further in the propaganda session, he had a short cell phone conversation, and he and his interpreter climbed into the SUV and drove off.

Now I wanted to see the damage to my airplane. First, we walked along the right wing to the number one engine.

"God, Senior, would you look at that," I said, touching the jagged, sheared-off end of one of the propeller blades. The alloy was hacked away for at least a foot so that the entire red-and-white prop-tip warning stripe was missing. On the prop blade above it, some of the stripe remained, but the metal was so chopped up, the once-balanced blade looked like one of the broken flint arrowheads we used to find as kids sticking out of the riverbank back home.

A third blade also lacked its tip and was amputated in a ragged stump. And there were lines of greenish paint streaking from the serrated edge. But the Finback had no green paint on its *outside* skin. The green had probably come from the fuel system or hydraulic lines deep in the interior. This was a stark visual demonstration of the monster chain-saw noise we had heard during the collision.

I again remembered the endless jolting vibration pounding the control yoke in my hands as I struggled to keep the plane flying. Our propellers were meant to be as finely balanced and synchronized as clockwork. Instead, we had flown with this clattering chaos at the end of the left wing.

At the blades' hub, the damage to three of the roots was just as chilling. The prop roots, made of a composite material even stronger than the alloy, were badly chewed up from debris that had exploded from the Finback. In fact, one of the prop roots had been gnawed almost in half. Just a little more damage, and that blade would have shattered, flown off, and probably slashed into our fuselage.

*How the hell did that blade stay on?* I thought, feeling again the relentless chatter of the windmilling engine. Then I glanced down at the hot concrete. A wide pool of oily hydraulic fluid had spread beneath the prop hub, evidence that the gears had been stripped, and the blades virtually welded in place, flat to the airstream. Senior reached up, trying to push the prop through a turn. But it took real effort just to budge it. Because of the stripped gears, he had to turn the whole turbine. I jumped back, worried that the shattered blade might fall off and hit me.

I walked under the wing, relieved to be in the shade after the oppressive weight of the sun. But, looking up, I saw a rectangular band of bright sunshine glaring through a rip in the aluminum skin of the right aileron. The hole was over a foot long and about four inches wide. Near it there was a black smear, probably from the anti-icing cuff on the leading edge of the Finback's vertical tail. Several small antennas had also been sheared off. Thank God that larger piece of debris had not ripped into that aileron. If it had, I would have never managed to regain wings-level flight.

We had also just escaped major damage to the number two engine or maybe punctures to the fuel tanks inside the wing. Or both.

I went to examine the dramatic damage to our nose. The left pitot tube was mangled. This confirmed my suspicion of why Jeff Vignery had a completely unreliable airspeed reading on his cockpit indicator. As I bent to look into the nose-wheel well, my hand came away discolored by red paint. Then I saw the wavy blood-red lines curving down the fuselage near the little oxygen filler inlet door. With a ghoulish feeling, I realized that red paint came from the Finback's large side number stenciled just beneath the fighter's cockpit. That was where the severed fuselage had struck us a fraction of a second before impacting our nose cone.

I stared into the gaping hollow once covered by that nose cone. There was no trace of the large weather radar antenna. Its mounting brackets were also gone, leaving behind at least four separate holes through the pressure bulkhead.

"That's where all the noise was coming from," Senior said.

I nodded grimly. There were also a number of ripped-out rivets and smaller fittings in the green aluminum skin of the bulkhead. Up at the top, the harness of the radar's cannon plugs that had flogged the right windscreen hung, now silent and unmoving.

Looking closer, I saw that the circular alloy nose ring, one of the strongest parts of the airframe, had been cracked and distorted and that the surrounding outer skin of the fuselage was peeled back and rippled like an accordion, almost to the windshield. And the right airspeed pitot tube was also distorted by the power of the collision. The strength of that nose ring had probably saved our lives. If the Finback had struck a few feet farther aft, the force of the impact would have been much more devastating.

When I stooped again to inspect the nose gear, I found the assembly in perfect condition. That had been my worst fear during that high-speed, no-flap landing. But *somehow* the wheel and its strut had survived the radiated force of the savage blow without springing a hydraulic leak.

"Take a look at number three," Senior called.

A chunk of prop blade about the size of a good dog bite was missing between the red-and-white tip and the black hub. Some piece of heavy debris, maybe the dish of the weather radar, had hit that prop hard enough to gouge that hole in the tough alloy. But by then we were already pitched up, snap rolling left with the right wing high. Otherwise, the debris cloud would have ripped the entire number three prop apart, and I doubt if we could have saved the aircraft. Or ourselves.

The Chinese had rolled a maintenance cart to the end of the left wing, and Senior Mellos simply climbed up onto it to look into the number-one engine.

"Do not touch anything," the base commander ordered.

Senior shrugged. "Well," Senior said, "number one definitely is fodded to shit."

This was no wonder, considering the degree of Foreign Object Damage the engine had suffered ingesting the explosion of debris from the Finback. On American military bases, people conducted daily FOD sweeps of the runways. And even a piece of foil off a cigarette pack was considered a hazard to a turbine engine turning at several thousand revolutions per minute. Somehow, we had survived a shower of FOD debris infinitely more threatening.

"You must come with us inside airplane," the base commander persisted.

"I do not need to see inside the aircraft," I answered stubbornly.

"You agreed to come inside airplane."

"No. I said no one would be allowed inside the airplane."

"You said you would come."

We stood in the baking sun, and salty sweat pooled in my hair and ran down my face and neck.

"I told you that I would *not* enter the airplane."

Incredibly, this haggling lasted almost forty minutes.

Then I turned to Senior Mellos. "Senior, do you want to go inside to see if anything's been messed with?"

"Roger that, sir."

That was another practical compromise. Senior could honestly answer he didn't know anything about the reconnaissance equipment in the backend. Almost his entire long career had been in P-3C Orions, not the EP-3E variants. So there was no danger of him revealing sensitive information. He and the base commander climbed the ladder, followed by two Chinese video crews.

I worked my way along the right wing, but the only damage appeared to be on the one prop blade on number three. The Big Look Radar's "M&M" dome was undamaged, as were the antenna covers farther aft.

Then I looked up at the tail. My mouth went dry even though I had just taken a deep swig from my water bottle. The long wire of the High Frequency radio antenna that ran from the top of the fuselage to the high tip of the tail fin had been ripped loose from the forward anchor point. Whatever had struck the antenna—probably our own nose cone sailing backward—had deflected the wire to the left just as we snap-rolled in the same direction. About forty feet of flailing antenna had wrapped around the left horizontal stabilizer of the tail assembly, jamming the elevator trim tab. That explained why I couldn't trim up the nose as I maneuvered low over the island, setting up for landing.

But as I stared up more intently at the tail, I also saw groups of popped rivets in the elevator itself. And I realized with a sick feeling that the metal was bowed out and distorted.

If I had tried to pull out of that dive too hard, we would have probably lost the whole elevator. The plane would have gone into a steep dive from which it could never have recovered.

I thought back to those frantic minutes after the collision. It was conceivable, I suppose, that a couple of people might have been able to bail out following a catastrophic malfunction in the tail. But I couldn't really see that happening. And I certainly would not have been able to ditch the plane or land it at Lingshui missing an elevator to control vertical movement.

Once I had regained wings level, but we had still been sinking rapidly, Johnny Comerford had managed to send off a brief secure radio message stating that the Finback had hit us and that we had called our Mayday. If we had then lost the left elevator, that probably would have been the last news America would have heard from Kilo Romeo 919.

I can't imagine how the U.S. government would have reacted. We had been sending regular position reports, so our chain of command knew we were in international airspace when the Finback began harassing us and the third clumsy pass ended in the collision. Chinese denials would have only worsened the situation. Obviously, I was glad we had survived. But I also recognized that no matter how long our detention dragged on, it was best for both countries that I had managed to regain control of the airplane and land at Lingshui.

Walking back toward the nose, I again scanned the readily visible damage, realizing the plane might have suffered internal airframe injuries that I couldn't see. There was no doubt that this tough old airplane had been subjected to G loads beyond its design tolerance. The people at Lockheed Aeronautics and Hamilton Standard who made the airframe and props had built a flying tank. We had survived a midair collision with a big, heavy fighter almost the same size and weight as the powerful American F-4 Phantom of the Vietnam War.

Standing in the sweltering heat of the Lingshui taxi ramp, I realized I was looking at a miracle.

Senior Mellos reappeared, and the three of us were taken to the guard van and driven back to the barracks. He was chuckling as he related the scene inside the plane.

"Why is this broken? Why is that broken?'" he said, mimicking the Chinese. "There were broken bits and pieces all over the deck. Wires yanked out and hanging down. I told them, 'Don't know. I was up in the flight station the whole time.'" Again Senior grinned. "Then they said, 'Make note of the damage.' I told them I sure would." Now Senior smiled broadly. "Lieutenant Comerford really did a Paul Bunyan job in there."

In the barracks, Pat reported that the Chinese had tried to take Petty Officer Shawn Coursen for interrogation. But he had refused. I was proud of him and realized the crew was holding up well to pressure.

When the crew assembled, I told them the extent of visible damage to the aircraft. Some of them looked troubled hearing this, as they fully realized for the first time just how close we had come to death. But I felt they had a right to know because they had behaved heroically in completing the Emergency Destruction plan while those of us in the flight station were fighting to save the airplane.

We were almost finished with lunch when several Mercedes and two small buses pulled up in front of the mess hall. A group of officers got out of the cars, and I was summoned to meet a new officer with three stars on his epaulettes who wore his hair in a laser-sharp flattop. Even though we were standing only a couple of feet apart, he shouted so that his voice echoed around the room. Poor Lieutenant Tony seemed cowed by this guy's tone and angry expression.

"These conditions are not up to your standards," the officer yelled. "In humanitarian interest, we are taking you to other quarters. You must get ready to leave."

Then, in the same tone, he bellowed a set of strict rules. The trip would last two and a half hours. For our "safety," we must not cause the buses to stop. We could not touch the curtains on the bus windows. And there would be absolutely no talking allowed.

"Where are we going?" I asked.

"Better quarters" was the only reply.

"That's not acceptable," I said. "We need to know."

"You will find out."

Now we began to haggle about the rules. I said we needed to be able to stop at least once and use a toilet.

"You cannot stop," the officer said. "We are concerned for your safety."

Instead, they agreed to give us empty water bottles to use on the bus. That would not help the women crew members very much, I said. But the prudish Chinese seemed mortified by this answer. I then insisted we be allowed to speak, but they said it would disturb the drivers. I said we would whisper, and they reluctantly agreed.

In early afternoon, we were lined up in front of our rooms and videotaped while an officer ran metal detectors over our flight suits and demanded of each crew member, "Do you have weapon?" Our only possessions were the meager toiletries they had issued us and some extra fruit the mess hall staff had offered after lunch. We were videotaped again as we climbed aboard the two small buses. More propaganda.

I had the three women on my bus and divided the other officers between the two vehicles, with Pat in charge of the second bus. Ahead of us, two SUVs with armed police escorted the little convoy. I sat in front with Lieutenant Tony across the aisle, and we engaged in casual conversation, despite the "no talking" rule supposedly still in effect. After pulling out of the base, we drove north along a narrow potholed road a short way, then joined up with a modern freeway that wound along the coast of the beautiful island. Through the front windshield, the view was spectacular. There were stands of palm trees and white beaches that matched anything I'd seen in Hawaii.

Johnny sat next to me, and we chatted about the chances of us actually meeting with American representatives before midnight as the base commander had promised.

"Maybe when we get there," Johnny said, "they'll just say 'You are free to go.'"

We laughed softly. "Maybe," I said. "But don't hold your breath."

About an hour north of Lingshui, we began seeing signs in both Chinese and English announcing the distance to Haikou, the principal city of Hainan Island. That was obviously our destination. An hour or so after that, we left the freeway and pulled into a downtown area where the streets were choked with small cars and three-wheel motorbikes laden with goods and produce.

We stopped at the steel-gated fence of a reasonably modern seven-floor hotel. There were armed guards at the gate. And the street itself had been cordoned off by more armed military.

"Upstairs," they ordered us, "quickly, quickly." This did not seem to be gratuitous bullying or an act. It looked to me like the Chinese were really worried about security. Probably their internal propaganda had worked too well, and there were a lot of people in this city who wanted to have a shot at us for "killing" their pilot.

"Rooms have been assigned," we were told as we climbed the stairs.

Pat, Jeff, and the three women crew members were led off at the fourth floor, while the rest of us were ordered to continue to the fifth.

"It is customary for the senior officer to have his own room," Lieutenant Tony translated for a man in casual civilian clothes who we were told was the "hotel manager."

They took me down a short side corridor from the landing. My room was large, with a separate sitting area, a bedroom alcove, and a bath. Lieutenant Tony's room was next door. There was a guard desk at the top of the stairs, blocking these two rooms from the corridor. The rest of the crew was assigned to rooms down the length of the hall. Senior Mellos had the first room on the right, which put him in the closest position to mine. It turned out that all the rooms had cable television hooked up to the Chinese networks, but that cable link had been shut down. When we turned on our sets, all we saw was a blue screen.

As soon as we were all settled, Johnny and Dick Payne came to see me and I asked them to pass the word for an "all hands" meeting in my room. But before they could, the manager appeared again and announced the new rules. "You are not allowed to leave your rooms or go up and down stairs," he said.

"How am I supposed to communicate with my crew?" I demanded.

"You cannot leave your room," he repeated. "You cannot go into any other room. You cannot go upstairs or downstairs," he said flatly. "Our leaders say these are the rules."

"Who is your leader?" I asked.

The manager smiled with embarrassment. I don't think anybody had ever asked him that question before.

"Our leaders say you will follow rules" was his only reply.

By now, a crowd was forming.

"Are we prisoners?" Johnny asked.

The manager scowled. "Does this look like a prison?" he said, nodding to the row of hotel room doors.

"No," Johnny said. "But we're being treated like prisoners."

"Why did the rules change?" I demanded.

"This is a new place," the manager said. "Here we have new rules." He turned and marched away.

I told Lieutenant Tony I needed to call a crew meeting. But when he returned from making this request, he simply said, "The leaders say it is impossible."

Twenty minutes later the guards came to take us downstairs to the small second-floor dining room. We all hoped that the promised representative of the American Embassy would be there. But the only people in the room were guards and interpreters. Four-chair tables stood in two rows, and a longer table served as a chow line.

The guards issued their usual orders, "Sit. Eat."

We ignored them and remained standing until everyone had food and we had said prayers. The guards up here did not seem impressed by our praying. But everyone continued to ignore their shouted commands, "Be quiet. No talking."

The food in Haikou was a cut above what we had been served in Lingshui. We got bread as well as rice, and there were a greater variety of vegetables. A couple of the dishes even had small bits of unrecognizable meat.

Back on the fifth floor, another argument ensued when I renewed the request for a crew meeting. Again the Chinese answered, "That is not possible." When I pressed them, they simply said their "leaders" would not give permission.

"Rest. Rest," they said, pointing toward the doors of our rooms.

Actually, that sounded pretty good. I was drained, having gotten a total of about four hours real sleep since takeoff at Kadena three days before. I also felt pretty grubby. But when I inspected the bathtub with its dripping handheld telephone shower, I was alarmed to find an open electrical box with exposed wires on the wall. Obviously, building codes were not too strict in the city of Haikou. Nevertheless, using the shower carefully and staying well clear of the electrical box, I did manage to get cleaned up without being electrocuted.

Then I reminded Lieutenant Tony that we had been promised we would see an American representative that night. He assured me everyone understood.

Around midnight, just as I felt I could actually sleep, Lieutenant Tony knocked on my door. "Get up, quickly," he said. The American representative had arrived.

*All right,* I thought. Finally. Now we would learn what was going on and how long the Chinese intended to keep us here. After the earlier mini-rebellion in the dining room, the crew was looking a little bleak because the Chinese had not backed down and had effectively confined us to our rooms and prevented us from talking across the corridor.

When we assembled in the dining room, the same officer with the piercing voice and the flattop haircut laid down the ground rules for the meeting. A new interpreter in an unfamiliar uniform and insignia translated as the officer read from a document. In the "spirit of humanitarianism," we would be allowed to see the U.S. representative. But we would not be allowed to pass him anything or to touch him. He would stand at a desk in front of the tables and a Chinese official would be in the room to monitor everything that happened.

Even with these strict rules, the crew buzzed with excitement. Some still hoped we were going home that night. I sat in front with Senior Mellos, Pat Honeck, and Johnny Comerford. Whenever the people behind us spoke above a whisper, the officer with the piercing voice yelled, "Quiet. Do not talk."

An hour passed. Then another twenty minutes. I was exhausted and nervous. *Now what the hell is going on*?

Finally the door opened. "They are coming," the officer screamed. "Sit straight. Get ready."

We saw the glare and shadows of a video lamp and the flash of still camera strobes. In the hubbub, I called to Johnny, who was closest to the door, "Let me know if it's a military rep."

But I didn't have to depend on a scout to identify the man who strode through the door. Army Brigadier General Neil Sealock, the U.S. Defense Attaché at our Beijing embassy, was tall and square-shouldered. The left breast of his green uniform blouse bore multiple rows of decorations and ribbons.

"Attention on deck!" I ordered the crew.

We rose in unison and snapped to attention.

"At ease," General Sealock said, his eyes scanning the room.

U.S. Consular Officer Ted Gomm followed General Sealock. A heavyset Chinese official sat silently off to the right.

"Take your seats," General Sealock said. "We've got to get this going quick because they gave me only forty minutes to see you. I need some information on paper, your places of birth and dates of birth."

As he raised his pen above his open notebook, the Chinese official shook his head. "No. That is against the rules."

General Sealock looked at the man in exasperation. "This is information *you* requested. You told me to get it. I'm just writing it down like we negotiated. I'm not passing anything."

"Ah . . . all right," the official conceded.

General Sealock looked evenly at the man, then down at his watch. "Now you're going to give us back those sixty seconds toward the forty minutes."

*We've got the right guy here working for us,* I realized. It was fantastic for morale to see the U.S. side standing up for us.

Once he had collected the personnel information, General Sealock rose to his full height and looked us all in the eye, one at a time. This was what is known as command presence. It was meant to inspire people, and I can certainly say it did so that night.

"Here's the situation," General Sealock said. "The incident is on the news around the world every ten seconds. Your family members have all been briefed and are being kept well informed. I have been requesting to see you since thirty minutes after you landed, and so far I have been granted only these forty minutes, and I had to negotiate all night for these visiting rules. But I'll try to see you every day. The country is solidly behind you. The president himself called me just before I came into this building and I'll call the president personally as soon as I leave here. How are you doing?"

"Morale is high, sir," I said.

"Great, Lieutenant," General Sealock said. Again he looked at his watch. "I've got time to take one personal message to send home from everybody," he said, flipping to a new page in his notebook. I was the last. I asked that he call my mom and that she spread word to the family and Roxanne.

"We're looking forward to going home, sir," I told him, "and we all have complete faith in the U.S. government."

"I'll be working until I get you home," General Sealock assured us.

"How are things going?" I asked.

"There are around-the-clock negotiations," he said, "but we don't have anything for you yet."

As General Sealock prepared to leave, he passed out a small Care package of Snickers and Australian Kit Kat bars. The forty minutes were over. Once more we came to attention as the general and the consular officer left the room.

Climbing the stairs toward our rooms, I sensed the crew's reenergized morale. Before the general's visit, we had no idea if the American people, or even our families, had been informed about the collision and our landing safely at Lingshui. It was possible that the entire incident would have been classified. But now we knew otherwise.

General Sealock had said the world's news media was focused on us. That scrutiny could only work in our favor. The Chinese were desperately eager to have Beijing named host city for the 2008 Olympic games. And the Chinese government badly wanted admission to the World Trade Organization. Detaining us indefinitely could erode their chances of gaining both goals. Unless, of course, they could pressure the crew or me into admitting we had caused the collision while spying on China.

Because of the Chinese official in the room, I had not mentioned to General Sealock the interrogators' threats of violence, reprisals against our families, or charges of espionage. But I felt much better now having reestablished contact with the U.S. government. Whatever the Chinese chose to do with us, it would have to unfold under the eye of the American Embassy, and the unblinking gaze of the world media.

So, physical force against us no longer remained a real option for them. Now we were locked in a battle of wills.

General Sealock had not tried to sugar-coat the status of the negotiations to make us feel better. He'd been frank and straightforward. I respected that in a leader. But back in my room, alone with my thoughts, I had to recognize the meaning of his report. The Chinese had not yet agreed to release us despite the round-the-clock discussions in Beijing and Washington. They were a people of intense national pride who would never accept the reality that their pilot had provoked the collision through a combination of inept flying and aggressiveness. But that was the simple truth. And I would never admit otherwise.

But how long would the Chinese keep up the pressure for me to do so, and how long could I withstand that pressure?

That night it became clear that they intended to keep me isolated from the crew and use sleep deprivation as their principal weapon to break my will. As I lay back trying to sleep, the guards outside would tap on the door every fifteen or twenty minutes, enter the room, turn on the lights, and stare at me. Once they used the pretext of coming in to check my water thermos and refill it. But usually they didn't even bother with an excuse.

At around 0230, I rose and stretched the kinks out of my limbs. I was still trying to operate on about four hours sleep in the four days since I had woken in the Habu Hilton Bachelor Officer Quarters at Kadena. My nerves were getting very bad, and I recognized the jangly onset of paranoia. Stuck out here in this side corridor I could do little to prevent the Chinese from yanking out one of the crew for intense interrogation. That was my greatest worry. I was confident I could handle anything they could throw my way. But I was not sure everyone else could take it.

That was the fear that haunted me as I lay on the bed, waiting for the guards to come in with their stinking cigarettes and flip the light switch yet again.

• • •

The next day at lunch, Pat asked me if I knew that a doctor had come that morning to treat one of the woman crew members who was still suffering a severe headache four days after landing.

"They sure as hell did *not* tell me," I said, furious.

I demanded a meeting with the Chinese officers. "I had someone in my crew sick and you did not tell me," I said, not bothering to hide my anger. "That is not acceptable."

I requested that we all be placed on the fifth floor. The crew on the fourth floor could carry beds up to my room, and the three female crew members could take it, while I could sleep on the third bed in Pat and Jeff's room. The hotel manager carefully wrote down the list of our requirements, which he promised to take to his leaders. By then we had learned from a shoeshine mitt that this was a military facility called the Nanhang No. 1 Guesthouse, and not a civilian hotel. So everyone we encountered here from the cooks to the manager was a member of the People's Liberation Army/Navy. And they had no reason to fulfill our requests. But I hoped to negotiate with them and reach another working compromise. "If we are allowed on the same floor and to speak freely," I said, "you have my word that no one will try to escape or leave the floor."

Again the manager made careful note of my words.

That afternoon, Jeff and Pat argued their way past the guards and came upstairs to see me. I explained the requests I had made and they asked if they could stage another hunger strike if we did not get a positive answer.

"That sounds good to me," I said. "Let's do it."

I also passed the word for everyone to wash their flight suit at the same time, so they would be dry to wear at the same time. We had to start looking a little more military. I know my own flight suit stank so badly of stale sweat that I almost gagged when I pulled it on.

At 1900 hours, the guards came to the fourth floor to call Pat, Jeff, and the women to dinner. As we had agreed, they refused to go because we had not yet received word on our earlier request. A few minutes later, the guards called me down, and I insisted on taking everyone from the fourth floor up to the fifth, where we could have a crew meeting. But before the meeting began, several officers and interpreters arrived and there was a confused, angry discussion in the corridor. I again assured the Chinese that no one would leave the floor if they allowed us to be together. But they refused and began shouting.

"Being together is as important as food to our health and morale," I said. "So, I don't think we will eat tonight."

The officers glared at me. There was real bitterness and anger in their eyes.

I looked down the corridor. "Is anybody hungry?"

"No, sir," the crew called.

I again told the Chinese we would not eat that night. They were practically twanging with pent-up rage.

When they had left, the crew again divvied up some fruit and candy and we went back to our rooms. But I was not left in peace very long. Several officers, accompanied by the usual video camera to harass me, stormed into the room.

"You must take the responsibility for any illness of your crew," Flattop said. "You must sign a statement."

"We are not going to eat until we can see the U.S. representative again, and also be able to talk freely among ourselves," I told them.

They marched out of the room. But at 2240 hours, the guards were back, ordering me to assemble the crew for a meeting in the dining room. When we got there, Flattop read from a printed statement in an angry harangue.

"You are not guests in the People's Republic of China," he said loudly.

"You are responsible for the death of one of our pilots. You are responsible for espionage against the Chinese people. You took our friendly gestures of feeding you as a sign of weakness. But if you decide to make threats, you do so at your own risk."

The food trays were set up at the chow line, and we could smell the steamed rice and vegetables. I certainly *was* hungry, and even the aroma of the mystery meat that some of the crew called "cat food" was tempting.

But none of the crew seemed impressed by the officer's statement. I turned to them. "Is anybody hungry?"

"No, sir."

On the way back up the stairs, I thought everyone seemed triumphant in their defiance. This was really becoming a battle of wills. Now it was a question of who would give in first.

After the bond of camaraderie in the dining hall, my room felt even more isolated. Throughout the night, the guards banged open the door, turned on my lights, and scraped furniture loudly across the tile floor. If I slept at all, it was in brief, troubled snatches.

Chapter Seven

---

# HAIKOU, HAINAN

## 4 April 2001, 0850 Hours

Flattop and the guards came to march us down for breakfast a few minutes late. I could see from Lieutenant Tony's manner that everyone was worried that we would continue our rebellious ways.

"We all go now for eating," Lieutenant Tony said, trying to sound reasonable.

I felt sorry for him. It was clear that he did not like these confrontations. Hell, he didn't even want to be here. The senior officers treated Lieutenant Gump and him badly. The two young interpreters hadn't even been given a change of clothes since they'd arrived at the mess hall in Lingshui the first day.

"What about the request that we gave the hotel manager last night?" I asked, looking right at Flattop. "Can we speak freely among ourselves and move from room to room?"

"Our leaders say this is not possible." Flattop's expression was bland, but his jaw was clenched.

"Have our requests even gone to your leaders?" I demanded. "We need to see General Sealock again about our requests."

"We will pass them," Flattop said. "Now you go to eat."

A video-camera man had been taping this exchange. Now he focused on me. "Is anybody hungry?" I called down the hall to the assembled crew. "How is everyone doing?"

"We're okay, sir," someone said.

"Not hungry, sir," another replied.

The camera was getting all this, and Flattop was beginning to lose his cool.

I turned to face him directly. "The crew would like to have Bibles."

He frowned in apparently genuine incomprehension when Lieutenant Gump translated this. Then he ordered me back to my room.

The confrontation in the corridor gave Pat Honeck the chance to move among the crew and speak with most of them. The three women all reported they were doing fine. A few of the others had minor health complaints, but everyone said they could stick with the hunger strike until our requests were granted.

Back in their room on the fourth floor, Pat and Jeff Vignery sat down to try to remember Biblical passages and the verses of hymns that we could use in religious services next time the Chinese allowed us to assemble. As hard as it was for them to comprehend, we had needs that went beyond physical hunger.

At 1300 hours, the Chinese told me to assemble the crew on the fifth floor once more. Again Flattop announced we would go downstairs for lunch. And again I asked him if our request had been granted.

"Our leaders say no talking, everyone must stay in his room," he said. Now his tone was openly angry.

"I will ask my crew if they are hungry," I said, taking the opportunity myself to move down the corridor, where people stood talking softly in small groups before their doors. This was the first chance I'd had in three days to speak directly to many of them, and I walked slowly.

"How're you holding up?" I asked. "You doing okay?"

Everyone reported they were still good to go on the hunger strike.

"My crew is not hungry," I told Flattop. "But we certainly would appreciate receiving something to read and some more decks of cards." Lieutenant Tony had managed to get us some cards in Lingshui, but now that everyone was confined to their rooms, there were not enough decks to go around.

"You make too many requests," Flattop said. "But you do not cooperate."

Before the Chinese ordered the crew back into their rooms, Pat Honeck took out a small piece of notepaper and read a Psalm. Then we bowed our heads while Jeff led us in a quiet prayer.

"No talking," Flattop shouted. "Inside rooms!"

"Lieutenant Osborn," he said, jabbing his finger toward me. "You must come now."

When the prayer ended, people filed silently into their rooms. I turned and followed Flattop.

As I had dreaded, he took me to another smoky room, where the Lingshui base commander and the admiral were waiting. The video cameras were already set up. Now no one tried to maintain a neutral expression. They were just plain angry at our refusing to eat.

"You cannot make threats," the admiral shouted. "For your own safety, you and your crew must cooperate with all rules."

"We were promised our requests would be answered," I said. "I must be able to see my crew, to speak with them," I insisted.

"This is not possible," the base commander said inflexibly. "No one can leave his room, for reasons of safety."

Incredibly, this same exchange spun in a slow, angry circle for several hours. I sat on the hard wooden chair, staring through the glare of the camera lamp at the men behind the desk as my legs went numb with pins and needles. I would not yield, and neither would the Chinese.

Finally, I tried to reach another practical compromise. "Is the crew safe in the dining room?" I asked tiredly.

"Of course," the admiral snapped.

"Well then," I said, "we could use our time at meals to discuss any necessary information."

"What information?" the base commander asked suspiciously.

"On our health, basically," I said.

The two officers briefly weighed my proposal.

"Lieutenant Osborn," the admiral said, "you now agree to eat?"

I had another card to play. "My crew cannot exchange information if the guards make us hurry at the meals. We must have as much time as necessary."

Again the two officers conferred.

Finally, the admiral nodded, "You will have the time. And you will no longer refuse to eat."

I played another card. "Can we see General Sealock soon?"

"If you eat."

"I agree," I said, returning his nod. "Did you receive our request for reading material?"

"After you eat," the admiral said, "it is possible that you will receive newspapers in the English language."

"I must now discuss this with my crew," I said. I had not won a complete victory, but I had the admiral's permission for us to extend our mealtimes and to speak in the dining hall.

At 1915 hours I assembled the crew on the fifth floor and explained that the Chinese had agreed to some of our conditions. "We can take as long as we want to eat and we can speak during meals," I said. "That's not as good as being able to move between rooms and speak freely up here, but it's the best I could get. What do you think?"

By now the people with minor health complaints like diarrhea were beginning to feel weak from the fast. And the Chinese had started cutting back on our drinking water, switching smaller bottles for the larger ones we had been given earlier. We had probably gotten the best deal we could from the admiral at this point. Everyone accepted the compromise.

"Let's go get some of that good chow," Senior Mellos said. "Maybe they brought us fish heads from Lingshui."

The first test of the new policy came fifteen minutes later when we stood at the tables, food before each person, and Johnny led the prayer. Now the guards did not shout at us to sit and eat. Instead, they gazed silently ahead, not conceding defeat. When everyone had eaten their fill, we were able to talk openly about the problems—both large and small—that we faced. The health concerns did not seem to be major. In fact, the woman crew member with the persistent headache was slowly improving. Since I had learned of her condition, I'd been worried she might have been injured following the collision. But today things were better.

And Senior Mellos had done his normal effective job of supporting the enlisted people. The watch system that he and Petty Officer Shawn Coursen had set up at Lingshui also worked well here in Haikou.

After the meeting broke up, Lieutenant Tony distributed more decks of cards, a box with heavily perfumed women's deodorant, and several battery-operated electric razors. The razors were not going to help us much, because we all were sporting five-day Beverly Hills stubble.

As we left the dining hall to return to our rooms, I had the feeling that we could outlast the Chinese in the battle of wills.

I had been back in my room about twenty minutes, trying to sleep, when Flattop arrived with two copies of *China Daily* English language newspapers. Others in the crew had also received copies. Although I was pleased to have something to read, it was immediately apparent that the newspaper was the official mouthpiece of the Chinese government. Printed on thin, easily crumpled newsprint, much of the *China Daily*'s April 3 and 4 editions were taken up with alleged news articles and "opinion" columns condemning the U.S. government, the U.S. military, and my crew.

It seemed as if the timing of Flattop's newspaper delivery was meant to keep me from sleeping yet again. I was just too keyed up to simply put the newspapers aside and lie down. The longest and most detailed article was an account of a press conference Chinese Foreign Ministry spokesman Zhu Bangzao gave in Beijing on April 3.

According to the Chinese government, our airplane had "approached the airspace over China's territorial waters off the city of Sanya to conduct surveillance." That was when the Chinese navy sent two F-8 fighters to follow and monitor the American plane. Spokesman Zhu said the fighters reached our airplane 104 kilometers "from the baseline of Chinese territorial waters." (Which, in fact, confirmed we were in international airspace.) All three aircraft were said to be flying a course of 110 degrees toward the southeast, with our plane to the right of the fighters. No mention was made that the original intercept had occurred on *our* right side.

"The U.S. plane suddenly veered at a wide angle toward the Chinese planes, which were closer to baseline of the Chinese side," the article continued, adding misinformation to contradiction about our original relative positions. "The U.S. plane's nose and left wing rammed the tail of one of the Chinese planes, causing it to lose control and plunge into the sea," Zhu told the news media.

This account had the EP-3E ramming the tail of one of the fighters, which would conveniently explain the damage our aircraft had sustained.

"The pilot Wang Wei parachuted from his stricken plane," the article continued, "while the other Chinese plane returned safely and landed at 9:23." Now I knew the name of the aggressive, poorly trained Chinese pilot who had nearly killed us and had almost certainly killed himself in the process. It gave me little satisfaction to place a name with the face of the man who had flown so closely beside our left wing, his oxygen mask unsnapped as he shouted silently at us and gestured angrily.

Despite a protracted search-and-rescue effort involving ships and aircraft that was continuing under the "repeated instructions" of Chinese President Jiang Zemin to proceed "at any cost," no trace of Wang Wei had been found.

The Chinese government again accused us of "veering and ramming the Chinese jet at a wide angle," which was a violation of flight rules. Spokesman Zhu also accused us of breaching the sovereignty of China's "exclusive economic zone," posing a "serious threat to the national security of China." The position of the Chinese government could not have been clearer, I thought. Then I read the next paragraph.

Citing the consensus reached between China and the U.S. to avoid risking military actions at sea, the Foreign Ministry noted: "When military airborne vehicles encounter each other in international airspace, both sides should properly observe the current international law and practices, and take into consideration the flight safety of the other side so as to avoid dangerous approaches and possible collisions."

*Roger that, Comrade Zhu*, I thought. These people had specifically cited how their pilots should have been flying, but had not. And now they had etched in stone the accusation that it was we who had initiated the "dangerous approaches."

I jumped up and threw the flimsy newspaper down on the bed. If Flattop's late delivery had indeed been designed to piss me off and prevent sleep, the ploy was working. But, despite my anger, I was drawn back to continue reading the long article.

Following the collision, Zhu had said, we had been guilty of "a further violation" of international and Chinese laws by our "gross encroachment upon China's sovereignty and territorial airspace." The Foreign Ministry had formally demanded that the U.S. government "bear full responsibility for the incident" and offer a complete explanation. Our landing the crippled aircraft at Lingshui had been an "illegal" act because we had not received Chinese approval to enter its national territory. Spokesman Zhu also claimed that we did not issue any distress call or request to land on Chinese territory, even though "facts show that after the collision, the U.S. plane had the time and technical ability to issue such a request." In other words, they had been all over the flight station and saw that the FM radios were working fine on the guard channel frequency of 243 MgHz—just as they were when we made our repeated Mayday calls and sent specific messages to Lingshui declaring an emergency and requesting permission to land there.

Zhu's third major point was more Alice in Wonderland logic. It was a "fact" that we had rammed the Chinese warplane. "After the collision, the front part of the nose of the U.S. jet dropped off, and the airscrew of its second left engine was deformed, evidence that the U.S. plane veered into, approached, and collided abruptly with the Chinese plane." Zhu added, "This cannot be denied."

Maybe not by the People's spokesman in Beijing. But he had gotten his "facts" a little scrambled. The blades of our number one prop showed paint tracings from inside the body of the Chinese jet, indicating the fighter had veered sharply *up* into our wing. Had our aircraft abruptly veered toward the Finback, the fighter could have easily either climbed away on afterburner or rolled off and descended to a safe altitude below us—*if* Wang Wei had been flying at a safe distance from our wing with his airplane under control.

Instead, the damage to the EP-3E provided stark evidence of exactly how the collision had occurred: The number one prop blades had spewed out debris as the Finback lurched up into the propeller disk; the bottom of our left wing had been riddled with that debris, and the severed nose of the Finback had slammed into ours, leaving behind a telltale smear of red paint from its side numbers before smashing loose our nose cone.

But, again I realized, there was not going to be any independent investigation. The article continued to call our airplane the "culprit aircraft." Zhu noted that given the "complexity of the incident, China needs sufficient time to make the investigation." I felt sick when I read those words. This was the official Party line.

I pictured the frantic scene on entering the hotel from the cordoned-off street. The Chinese might actually have legitimate fears for our safety. If people across China had been bombarded with such lies and propaganda for the past five days, there might well be a lynch-mob mentality brewing on Hainan. I suddenly remembered the TV images of the mobs that had attacked the American Embassy in Beijing in May 1999 after a U.S. aircraft on a NATO bombing mission over Serbia accidentally bombed the Chinese Embassy in Belgrade, killing three Chinese journalists.

Then, the U.S. had correctly insisted that the tragic bombing had been accidental, the result of a target-selection error.

Now the Chinese seemed determined to squeeze us into admitting that our crew had caused the collision with the Finback.

During my early interrogations, I had hoped I was just facing zealous local officers like the base commander and the admiral, who were trying to pressure me into an admission of guilt so that they could hide the reckless conduct of the Finback pilots from Lingshui. But now I had to accept the fact that the Chinese government itself seemed determined to keep us here until my crew and I accepted full blame for causing the collision by "ramming" the Finback.

That was not the worst of it. The "most urgent matter for the U.S. side," according to Zhu, was for the United States to thoroughly review the incident, and then to "apologize to the Chinese side." Reading that line, I knew the thrust of the coming interrogation would be for the crew and me to make a formal apology. But an apology would mean that we had caused the incident, not the Chinese pilot, Wang Wei. If we were forced to make such an apology, we would be conceding that the false Chinese version of the collision was true. This was not acceptable. I would never apologize, and I would do my best to persuade the crew not to do so.

Spokesman Zhu's distortions continued. China, he said, "lost no time in allowing U.S. diplomats" to meet with the crew of the U.S. plane, "after taking into consideration humanitarian concerns," which, he added, demonstrated "China's sincerity and humanitarian spirit in handling the issue." That was just not true. General Sealock had told us it took him three days of nonstop negotiations to get to see us for forty minutes.

The long article closed with the Chinese government's demand that the United States "apologize to the Chinese side and shoulder all the responsibility arising from the incident."

I felt like ripping the *China Daily* into shreds, but I knew the Chinese had distributed only a few copies, and they had to be shared among the crew.

*This is going to be a longer battle than I thought*, I realized.

The familiar harassment routine continued that night. Whenever I slipped into a light sleep, the guards would enter the room, bang the furniture around, or prod me awake. They were little guys. I was becoming so enraged with their wake-ups that I actually considered snapping the neck of one guard and hiding him among the blankets in the closet. He had woken me four times that night.

But I quickly returned to rational thought. Such action was not in the best long-term interests of my crew. Resorting to violence would only confirm the lies in the Chinese press.

Still, I was rapidly going nuts from the cumulative lack of sleep and isolation. I found some satisfaction in writing "letters" to my loved ones, printing as well as my groggy mind allowed in the small pages of my notebook. I had no idea when and if these letters would ever be delivered, but hoped I would have the chance to do so myself, soon.

Writing to Roxanne, I stated honestly, "Things aren't good here, as you know. I'm confined all by myself and have no idea what will happen." I gave her my love and asked her to tell my dad that I was handling this detention in the way he expected I would.

In my next letter to Roxanne, I conceded, "It's kind of sinking in that I may be here awhile." Solitary confinement, I said, was very difficult. "I have nothing but my thoughts." But I asked her not to worry about me because I was tough enough to handle it and begged her to stay in touch with my mother.

In my letter to my mom, I said I knew how hard this situation was on her and the rest of the family. "I did what I had to do to save twenty-four lives. The fact that I am writing this letter to you today is a miracle from God Himself. Please know that I am strong, clear-minded, and *will* bring twenty-four men and women back to their families, where they belong."

Writing to my squadron skipper, Commander Bernard Lessard, I told him, "I am honored to have twenty-three of America's best men and women on my crew. Everyone has performed bravely and beyond the highest expectations. I will deliver you twenty-three of the finest airmen the Navy has ever seen. Our spirit is good, and everyone is of sound body and mind. We have done *everything* right and will return with honor. We have faith in God and in the United States, and no one doubts our safe return."

When the crew assembled on the fifth floor for breakfast, on the morning of April 6, Pat Honeck and Jeff Vignery crept up the stairs to present a brief skit, one of several they had written, to lighten the mood of our detention. This one was a takeoff on the television program *Crocodile Hunter.* Using his best Aussie accent, Jeff stalked around Pat, who was writhing on the floor.

"Ah," Jeff said, pointing down, "there's the beast."

Pat rolled around, roaring. The guards looked at us like we were insane. "Time to eat. You go downstairs."

We didn't respond, but instead watched the skit.

Jeff was now instructing us on the best way to handle a "big 'un" like Pat. "You scratch their belly, mates, and that tames 'em."

"March in line," the guards demanded. But, still laughing, we lollygagged down the stairs like a bunch of high school kids. It felt great to laugh out loud. Letting the Chinese know we wouldn't be so easily tamed as the crocodile.

After breakfast I spoke to the crew about the need to stay in good physical shape because we didn't know the possible duration of our detention. We had already devised ways of working out, using the scant furniture in our rooms as weights or crude exercise apparatus. Senior Mellos and Sergeant Pray had even filled their plastic washbasins with water and stood in the shower lifting them in multiple sets of arm curls. Again, I was feeling punchy due to the harassment of the guards all night. My nerves were jangly, and as we talked, I felt an almost physical hunger to work out to the point of exhaustion so that I could finally sleep.

But then the admiral, his two interpreters, and the regular video cameramen entered, trailed by two men in sport shirts and slacks whom we had never seen before. One was from the People's Liberation Army/Navy, and the other was apparently some civilian official.

The civilian and the new military officer took turns reading from an endless document that was basically a lot of convoluted language supporting a straightforward threat: If all the crew did not "cooperate fully" with the investigation, we would place ourselves in "serious" legal jeopardy. The officials droned on, citing the United Nations Convention on the Law of the Sea, and other international conventions and treaties to prove that we had violated Chinese airspace, caused the fatal collision, engaged in espionage, and landed on Chinese territory without permission.

"You face a very long trial," the civilian official said. "You illegally entered our country. You engaged in espionage and spying against the Chinese people. You must admit this, everyone."

As he spoke, I looked around and noted the expressions of concern among the crew members. Some of them were obviously scared and upset. They hadn't heard this kind of propaganda before. But I was used to it from the long interrogations.

"If you refuse to cooperate with the military investigation," the official continued, "you must all face a long criminal trial much later. It is much better to cooperate now so that you can return to your families. We will treat you with humanitarian spirit if you cooperate. If not, you must be treated as criminals."

The People's Liberation Army/Navy officer took up the harangue, basically making the same accusations and threats. Even though the admiral's interpreters were much less adept than Lieutenants Gump and Tony, we all understood what was being said.

Sitting beside me at the front table, Johnny Comerford was becoming increasingly agitated by the drumbeat propaganda. "Let's just get up and leave," he muttered. "We don't have to take this crap."

"No," I said. "Wait, they're just getting warmed up. They'll get to their point soon."

As the two speakers read on, and the interpreters stumbled through their clumsy translations, I turned back to face the crew several times, grinned, and made a grotesque face that I hoped conveyed the appropriate "fuck them" expression I had in mind. It must have worked, because the crew grinned back.

Finally, after two hours, the admiral spoke directly to me, but his interpreter's words were loud enough for everyone to hear.

"We must speak to each crew member alone," he said, looking around the room at the somber faces. "Everyone must cooperate with the investigation. We need simple information."

"We are not ready to give any additional information," I said.

"Only simple information," the admiral insisted.

"Give me some paper," I said. "I will write down everyone's name, rank, and Social Security number. That is the information we are willing to give."

"No," the admiral said, shaking his head. "We need more information. I think the others want to talk, but you are not letting them."

"They are free to talk if they want to," I said, speaking loudly enough to be sure everyone in the dining hall heard me clearly. "But I want to be there if they do."

The admiral lost his cool. "We are not afraid of the United States," he said, his voice rising to a shout. "We defeated you in Korea and in Vietnam. We must have full investigation. You will not return until you cooperate."

"I can give you their names," I said, "or they can talk to you if I am there."

"This is not possible," the admiral said. "Perhaps someone wants to talk but you do not let them."

I did not reply. Did they seriously think that I would break down and sell out my crew right in front of them? I guess they did think I had been weakened by the long hours of sleepless isolation. Well, being here with the crew made me stronger. The longer we could drag out these group sessions, the more strength I would gain. It was my job to distract the Chinese, to be the lightning rod and deflect them away from the crew.

I ignored the admiral's comment. "When do we get to see General Sealock and Mr. Gomm again?" I asked. "We haven't seen them in two days. Also, what has happened to our request to move among our rooms and speak freely?"

The admiral answered in a stern tone. "You are not visitors here," he said, glowering at the crew and me. "You are not welcome here. You are spies. You can be tried. Do you want to go to the criminal court and not cooperate with the military investigation? That court would not be good to you."

It was very tense in the room, but I looked to my left and saw Senior staring down at the table, on the verge of laughing. In the middle of the admiral's tirade, I had reached under the table and rubbed Senior's knee like a Hong Kong bar girl when the fleet was in. I had to do *something* to lighten the mood. I wanted him and everyone nearby to know that I wasn't intimidated and that I was just as nuts as he was.

Now the admiral demanded to remove certain backenders for individual questioning.

"No one can be questioned alone," I said. "I must be present."

"No, each person alone."

"I must be present."

The argument ran in circles until 1300 hours. We had been sitting in our chairs since breakfast, so they brought in lunch.

After the meal, the same bickering argument resumed, with the admiral still insisting that crew members submit to individual interrogation. I continued to refuse.

The tag team of the military propaganda officer and the civilian official took over, throwing out their accusations and threats. This session lasted three hours, and then we were told we could have an hour's rest in our rooms.

I was just getting into a series of arm curls using the folding chair, when the guards came to fetch me for interrogation.

The Lingshui base commander and Flattop were waiting for me in the interrogation room.

"Our admiral wants you to cooperate with us so that we can resolve the problem," the base commander said. "We want to go home to our families, too. But we can't allow you to see the American representatives until you cooperate with the investigation."

"We have cooperated," I said.

"Everyone needs a chance to talk, to help the investigation," Flattop added. "But you will not let them."

I realized this was going to be another tough argument, but I knew how important it was to protect the crew as long as I could. To throw the officers off, I nodded, as if in agreement. "They can all talk. But they do *not* need to talk alone."

"You prevent them from talking," the base commander said.

I shook my head. I could be as patient as they could. "If they want to talk alone, they can. But they do not want to. I spoke to them after lunch."

Once more we engaged in our unpleasant little debating society. I had reached that point of exhaustion where I was practically hallucinating. But I was not going to give in so easily. An hour passed, then two.

Finally, the base commander said in frustration, "What do you want so we complete the investigation?"

He was willing to negotiate. "The crew will answer questions in their rooms, two at a time." I forced myself to think clearly. "But the doors must be open, and I must be nearby. The crew can tell you their name, rank, date of birth, and Social Security number. But they will not answer other questions unless they choose to do so."

The base commander had been writing in his notebook as the interpreter translated. "I understand," the officer said. "Now you begin to cooperate."

I held up my hand. "There is more. No videotaping or recording of the questioning. And the interviews must be limited to ten minutes unless the crew members request to speak longer."

The base commander scanned his notes. "It is impossible for you to be with the crew because they will not speak freely with you present."

Again I shook my head, which I now understood was a very disrespectful gesture. *What the hell, let them think I'm a barbarian.*

"These are the only acceptable conditions," I said. "I do not want anyone to be intimidated."

"You are a very stubborn man, Lieutenant Osborn," the base commander said, looking down at his watch. It was almost 1730 hours. "But you care for you crew. However, we are missing one of our comrades and we must conduct this investigation. You may sit at the end of the hall and observe the questioning to note the time."

Apparently I had won another small victory.

But when I went to inform Johnny and the flight station crew on the new ground rules for interrogation that I had just negotiated with the Chinese, I found two sets of interrogators, with cameramen, were already at work behind the closed doors of two of the enlisted crew members' rooms. Another interrogation team with cameraman was just entering a third room. I strode down the hall and beat on one of the closed doors.

"What is this?" I demanded, speaking loudly to the base commander following me. "This was not in our agreement."

He spoke in rapid Chinese and the room doors opened.

"The cameras must go out," I said even louder. Doors popped open up and down the hall.

After another burst of Chinese, the cameramen left.

I also discovered that the Chinese were conducting the interrogation using printouts of questions with blank spots for the answers to be filled in. The questions included the crew members' home addresses and telephone numbers and those of their parents, the schools we had attended, and our specific assignments on the airplane.

"If you really want them to have this information," I told the crew members gathered at their open doors, "go ahead and give it."

They just grinned back. Obviously, people were not going to voluntarily answer those types of questions.

I observed the interrogations from the end of the corridor. Even though the enlisted crew members were questioned longer than the terms of my agreement with the base commander, they later told me they had not given in to the pressure for greater "cooperation," and had instead insisted they knew little or nothing about the actual collision. They had also resisted discussing their specific duties on board the aircraft, and simply stated "technician" when asked about their job.

When that questioning session ended, Flattop and the hotel manager brought me a larger Care package from General Sealock. The materials were in several cardboard boxes, which they wanted me to divide and distribute into small plastic bags for the crew. I was happy to see enough playing cards, so that each room would have a deck, and I didn't mind keeping one back for myself.

The powdered detergent was also welcome, as were sets of underwear, shorts, and T-shirts. The socks were short and very thin, but they were a lot better than what we had been wearing for six days.

When I came to the feminine hygiene products, I decided to go ahead and pull the chains of the two Chinese. "Hmm," I said, holding up a box of tampons. "What is this?" The two men pursed their lips and looked away in embarrassment. One more little victory in retaining my sense of autonomy.

Before dinner that night, the crew joined hands and we sang the Navy hymn, "Eternal Father." The words seemed very appropriate.

*Eternal Father, strong to save,*
*Whose arm hath bound the restless wave . . .*
*Oh, hear us when we cry to Thee*
*For those at peril on the sea.*

*Eternal Father, grant our cry*
*For those who fly up in the sky . . .*

I think we all felt protected since the moment of the collision, and many who had not been especially religious before were now becoming so. The old adage about no atheists in foxholes was certainly holding up.

At 2030 hours, General Sealock came to the dining hall for his second visit with the crew. Again he instilled confidence in us by his appearance and manner. And it was an encouraging sign that no Chinese official accompanied him.

Pat and Jeff had prepared a brief written list summarizing the exact conditions and sequence of events before, during, and after the collision with the Finback.

"I think Lieutenant Honeck has something important for you, sir," I told General Sealock. I wasn't more specific because it was certainly possible the room was bugged.

General Sealock nodded and opened his notebook as Pat began to read from the list.

"PB-20N," Pat said. "Solid." He was referring to the fact that our autopilot had been holding us on an unwavering heading before the Finback had flown into our prop. General Sealock made a note.

"INS," Pat continued. "Tight." Our inertial navigation had given us a tight position readout squarely in international airspace.

"Course zero seven zero," Pat said, "at two two five. One-eighty knots, steady." We had been flying northeast at 22,500 feet at a steady airspeed of 180 knots.

General Sealock wrote quickly.

"Left turn after impact," Pat continued. "Number one prop gone. Nose gone. Pressure bulkhead compromised."

I took up the list to provide information on the damaged items I had personally observed on the Lingshui taxi ramp. "Airspeed indicators, gone. Left aileron, large hole. Nose ring, trashed. Elevator, bent. HF wire, separated and wrapped around tail."

General Sealock jotted notes as fast as I spoke. For a moment I was silent, and we looked at each other. "Checklists, completed." He grinned, understanding that we had completed the Emergency Destruction plan.

"We got off several Maydays," I added, "and we called Lingshui directly on 243 declaring an emergency. No response that we heard."

General Sealock nodded. This was his first chance to hear a detailed account from our crew.

Now I described the landing. "We crossed the runway below 1,000 feet to make sure it was clear. I set up for a fast no-flap landing, and we touched down right on the hash marks. Rollout almost normal, considering the damage to the aircraft. That's pretty much it, sir."

"That's great," General Sealock said. "You obviously are doing fine, but I want to let you know that you're not POWs. I don't want you to say anything to the Chinese outside your squadron SOPs, or discuss things with them that make you uncomfortable. But you need to talk to these people so that we can move things along in the negotiations."

I nodded. "Yes, sir." This was the first sense of official direction I had received concerning the interrogation of the crew. Navy Standard Operating Procedures did not permit us to discuss classified information with the Chinese, and I was confident that none of the crew would do so. General Sealock's mention of moving the negotiations forward certainly sounded good to me. But we had to keep in mind his caution not to discuss anything that made us uncomfortable.

"I don't want to raise your hopes too much," General Sealock continued. "But remember, you are on everyone's mind back home. Public support is phenomenal. Don't worry. You are not forgotten."

"We know that, sir," I assured him. "We do have the faith. Don't worry about us. Just do what you have to do to get us home *right*."

"I can tell you here and now," the general said, gazing slowly around the room to meet our eyes, "the United States government will *not* apologize."

"Neither will we, sir," I said in the same tone. "We have nothing to apologize for."

Before he left that night, the general wrote down a brief e-mail message from everyone to be sent to our families and loved ones.

I asked him to tell my mom and dad that I was in good health and confident of returning soon.

That night the guards played their stupid tricks again, checking my water and adjusting my air conditioner every fifteen minutes. I was used to that. But what really got to me was the new guard at the desk outside, who spent the night hawking and coughing loudly and either spitting or swallowing phlegm. Every time he did, I felt like someone was scratching a blackboard.

But I actually had managed to slip into a light sleep, when the door banged open and a guard, accompanied by a sheepish lieutenant, delivered several copies of the *China Daily*. "These have just arrived," Lieutenant Gump said tiredly.

There were more tirades against the U.S. One front-page statement quoted President Jiang Zemin, who was about to depart on a tour of Latin America, as demanding that the United States apologize to the "Chinese people" for causing the collision and "bear all responsibilities."

This issue was plastered with similar statements from high Chinese officials, and included a statement from the Chinese Ministry of National Defense expressing "indignation and condemnation" that the "U.S. plane violated flight rules and unlawfully intruded into Chinese territorial airspace in landing at a Chinese airport." Over and over again, the news articles and editorials—it was hard to distinguish between them—hammered on the same theme. "The U.S. spy plane rammed and destroyed the Chinese plane and illegally entered Chinese territory."

The most recent edition, dated April 6, had a front-page photo of Wang Wei's grieving parents, with appropriate quotes edited to evoke sympathy and fire up maximum indignation against my crew and the United States.

But the most interesting article was a column on the left side of the front page. It cited another Foreign Ministry spokesman, Sun Yuxi, as noting that on Wednesday, Secretary of State Colin Powell had said that the United States "regrets that the Chinese plane did not get down safely" and "regrets the loss of the life of that Chinese pilot." But spokesman Sun said this was not enough. "What I want to stress is that the United States must assume total responsibility and make a full apology." But again there was absolutely no mention that the Chinese fighters had been harassing us at an unsafe distance. The Chinese government had not budged from its demands that the United States accept complete responsibility for the collision and deliver a "full apology."

But General Sealock had made it clear that this would never happen. I stared down at the smudged newsprint. If the Chinese could not wring an apology out of the U.S. government, they would concentrate even harder on squeezing us.

The next morning's skit was a send-up of *People's Court,* which Jeff and Pat called "Senior's Court." This session involved Jeff accusing Regina of sneaking an extra helping of shrimp from last night's dinner before everyone had received one. We were all excited about finally getting recognizable food and "the evidence" stated that Regina had become carried away and couldn't control herself. Senior judged her guilty and ordered her to go last in line like I did at every meal to ensure the crew would be properly fed.

When we finished eating that morning, the admiral spoke to me in a cold tone. He was still angry about our earlier open defiance. "I see that you eat now. But, if you are not hungry again, I will send my cooks home. They are tired. They need rest, and they do not want to see their food wasted."

Then he introduced another senior official from the People's Liberation Army, who launched into a tirade that seemed an almost-verbatim repetition of the Ministry of Defense position on the collision that I had read in the *China Daily* the night before. The official ended this long-winded session by demanding that we "cooperate" with the investigation.

I conferred with my officers and Senior, then told the Chinese officer, "I will allow the other crew members to be questioned. However, they will answer only the questions they feel comfortable with."

The officer seemed satisfied. He consulted a list and requested that Jeff Vignery, Ensign Richard Bensing, and Petty Officer Jason Hanser remain behind for questioning.

I stayed seated as everyone else rose to leave. Flattop motioned for me to also leave with the crew. I shook my head. "The rules are that I stay for the questioning."

"Those were the rules yesterday," he said stubbornly. "Today we have new rules."

"We'll be okay, Shane," Jeff assured me.

He was right, of course. The general had told us to be more flexible, and I knew no one would betray the trust the Navy had placed in him. I left the room with the crew.

At lunch, Jeff said the Chinese had set up two model airplanes—one of a Finback, the other of a big, four-engine Russian-designed Antonov-

124 jet transport—for the crew members to demonstrate how the accident had occurred. But they refused to do so, especially because the session was being videotaped, and the models were not of the same scale as the fighter was the size of the transport—the world's largest airplane. This was another breach of the rules the admiral had accepted.

That afternoon, the admiral and Flattop tried to pull the same stunt with me in the interrogation room. The usual video camera was set up, lamp blazing. But I also refused to touch the models. I knew the videotape could easily be edited to indicate my describing how our aircraft had "abruptly" turned into the innocent Finback. This was an acrimonious session, even though we didn't actually say much.

The basic exchanges were between the admiral and me. "You show us with these airplanes what happened," he would say over and over again. "How did you ram our fighter?"

"I cannot show you," I would say, "because we did not ram your fighter. Your pilot flew into our number one engine."

At lunch I had been given a copy of the *China Daily* quoting the second Finback pilot, one Zhao Yu, who claimed the two fighters had been flying "about 400 meters from the U.S. plane" when we turned abruptly left and rammed his wingman, Wang Wei's fighter. Zhao was, in fact, the witness I had wondered about.

While I was being interrogated, another team in the dining hall was questioning the women crew members. That night at dinner they reported they had stuck to the basic factual account of how the collision had occurred.

Earlier that day, Senior had cautioned the enlisted people about the shifting circumstances we now faced. The Chinese were rapidly changing rules on us in an effort to sow confusion. "Look," Senior said in a tone that seems unique to chief petty officers in the U.S. Navy, "don't get lazy around the guards. We might still be here a long time. This detention crap can be drawn out much longer than any of us want. But we're Americans, and we're sailors. So we'll watch what we say and what we do."

The isolation and tense boredom were definitely getting to me. I didn't want to read the *China Dailys* that the guards brought. I had no one to talk to. Sometimes, I tried singing. The lyrics to an old rock song I had played a lot as a teenager filled my mind. But I remembered only one stanza, and I'm not even sure it was accurate.

"Lord," I droned, "I'm wasted and I can't find my way home . . ."

The words I tried to fit into the Crosby, Stills, and Nash tune "Southern Cross" that I'd recently heard again on Jimmy Buffett's latest live album concert were wrong. So mangling the song, I repeatedly chanted like a madman, "Think how many times that I have fallen / Spirits are using me / larger voices callin'," while pacing around the room, hoping to tire myself out enough to sleep. The biggest fear that kept me awake was the thought of one of the other crew members being yanked out of his or her room and me sleeping through it. Therefore I was always easily awakened by the slightest noise.

When the tactic of singing didn't work, I turned to the deck of cards. I had almost never played, and didn't even know how to set up solitaire. The only game I was familiar with was king's corner, which my great-grandmother and grandmother had taught me as a kid in Nebraska.

They were hardworking farming women who endured harsh lives but took comfort in church and the innocent diversion of an occasional card game. My mom's mother, Dorothy Lichtenfeld, had died in February in a hospital bed set up in Mom's living room. I had been with her on emergency leave, holding her hand when she passed away, just as I had been when my great-grandma Clara Kallweit had died when I was nine.

I had felt Grandma Dorothy's presence as I began this detachment.

In fact, I was amazed and pleased that Pat wanted to name the plastic hula doll we had picked up in Hawaii on the way over "Dorothy." We mounted her on the center console of the flight station. She had been the plane's good luck charm. And it seemed to me that talisman had been pretty effective.

I sat on the bed and dealt out the cards for King's Corners, my hand on the right, Grandma's in the center, and Great-Grandma's on the left. Hour after hour, I lay the cards down, playing each hand in turn. I had been taught it was traditional for the winning hand to knock, so I rapped the headboard whenever one of us won. It would have been easy for me to cheat, of course, but I didn't. These were real games to me. I felt the presence of those strong old farming women supporting me, just as they had when I'd been a child.

I spent hours remembering the summers spent at their houses, attending Bible school at the tiny Zion Lutheran Church just down the gravel road from Grandma's. I daydreamed of the times when my cousins Junior, Richard, and I would play all day and night in the shelterbelt of cottonwoods near the farm, building multilevel forts from the pile of planks from a demolished barn. For a brief moment I had left that horrible hotel in China and returned to my people, the extended family that was my virtual tribe, on the farm in Nebraska.

But when the memories, the crazy singing, and the card games did not work, I turned to prayer. I prayed for the crew and for my family and to remain mentally strong.

"Lord," I prayed, "give me the strength to stay clear-minded, to retain composure. Give me guidance and influence my actions."

After that afternoon's interrogation, the Lingshui base commander and Flattop had come to my room.

"I want to go back to Beijing," Flattop told me with apparent frankness, revealing for the first time that he came from the capital. "We all want to go home just like you. What do you think of the admiral's words?"

"My aircraft did not cause the accident," I told them.

"We want you to cooperate," the base commander said in a reasonable tone. "We want you to go home. We want friendship between our two countries. But you must speak to us honestly."

"I have spoken honestly," I said, feeling the wheel of the endless argument grinding me down.

Flattop shook his head. "No. We have brought specialists from Beijing to examine the airplane. We are certain you turned left and rammed our fighter from behind. We know this because there is red paint on the front of your propeller and not on the back."

"You cannot dispute this now," the base commander persisted.

I realized that neither of these men understood much about aviation. The traces of red paint on the shattered ends of the props were from the red-and-white blade-tip warning stripes, and that paint was on both sides. They did not mention the smeared green paint that came from the Finback's interior and revealed how fast the out-of-control fighter had risen into the spinning disk of the number one propeller.

"Your pilot struck my airplane," I said. "I have already explained this."

"You must apologize," Flattop said.

"No, I will not."

"Why is it so hard for Americans to apologize when they are wrong?" the base commander demanded.

"I have nothing to apologize for," I insisted.

The base commander rose to leave. "Americans are trying to take over the world with your arrogance," he said in disgust.

• • •

At 0150 hours on April 8, the Chinese woke me just after I had fallen asleep to announce that General Sealock had arrived and that he would be allowed to meet with eight crew members. Beside the section-head officers and Senior, I asked that Seaman Bradford Borland join us. He was already awake because he'd just come off watch.

General Sealock said he'd had to argue all night for the chance to see us. I saw that he was becoming just as angry and frustrated at these Chinese tactics as I was. But he had succeeded in wearing down the Chinese. And I was certainly glad he had persevered. He had brought a batch of e-mails, including one from my mom and dad, as well as another Care package with candy bars and beef jerky.

"How are you holding up?" he asked.

"Morale is good," I said. "We're doing fine, sir."

Then Seaman Borland raised his hand. "Sir," he asked me, "can I ask a question."

"Sure."

"General Sealock," Brad Borland said with a perfectly straight face, "do you have any idea what the per diem rate is here on Hainan Island?"

We all broke up laughing. If the General needed any demonstration of our morale, he'd just heard it.

As he rose to leave, I asked General Sealock the status of negotiations.

"Nothing is resolved yet, Lieutenant," he said. "Everything is still up in the air."

Climbing the staircase, we were all quiet, each person absorbed in his thoughts.

Chapter Eight

# HAIKOU, HAINAN

## 8 April 2001, 0930 Hours

The day had begun well with another hilarious skit, in which Jefe walked the runway, aping a Paris fashion model, while Pat provided the commentary. In the dining hall, Pat recited Psalm 121, "I will lift mine eyes unto the hills, whence cometh my help . . ." We ended the religious service by singing several verses of "Amazing Grace." The Chinese had never given us the Bibles we had requested, but Pat and Jeff were doing a great job writing down the scripture and hymns they could remember.

After eating, I spoke to the crew about the interrogation they were undergoing. Unfortunately, I had become the local expert on the subject.

"It's usually just better to listen than to speak," I advised them.

Before we left the dining hall, Pat and Jeff distributed the Snickers bars and beef jerky that General Sealock had given us the night before. These little gifts were a real morale booster.

Now the Chinese openly disregarded the ground rules on interrogation I had established with the base commander and the admiral. Officers were being kept for questioning longer than the ten minutes I had agreed to, and the interrogations were being videotaped in a corner of the dining hall. But I was extremely proud of them. Although they appeared to cooperate in small ways, nobody gave the Chinese anything. And everyone stubbornly refused to accept blame for the collision or to apologize.

For me, the worst aspect of this period of detention was the continued isolation from the crew. I passed that morning in prayer, singing my nonsense lyrics, and playing king's corner with Great-Grandma and Grandma.

But I felt the overpowering need to be in closer contact with the crew. At lunch that day, I got together with Pat, Jeff, and Senior and we decided to hold a military awards ceremony that night to recognize the Sailor and Petty Officer of the Week. This was a Navy tradition in each squadron or division aboard a ship. The choice for petty officer was obvious. Shawn Coursen had emerged as a real leader of the enlisted crew members. The watch system he had devised at Lingshui still functioned around the clock and gave us all a sense of security, knowing that the Chinese could not remove anyone for questioning without our knowledge. At twenty, Seaman Jeremy Crandall was the youngest member of the crew. He had never given in to the Chinese threats and intimidation and always maintained his sense of humor and positive spirit. He was, in fact, an inspiration to several older crew members.

That afternoon, Senior Mellos wrote out the paper citations that we would present that night.

After dinner, we set up a table out in front and called the crew to attention. Pat took his place as executive officer, and I assumed the traditional position of the commander, several paces back, while Senior Mellos stood at my side. Pat would read the citations, and the men being awarded would come to me, salute, and Senior would present the citations.

"Seaman Jeremy Crandall," Senior called in his best quarterdeck voice, "front and center."

Pat read the citation, noting that this temporary paper would be replaced by a permanent official citation once we had returned to the United States. Crandall marched up proudly and we exchanged salutes. I shook Crandall's hand, and he beamed with pride.

Next came Petty Officer Shawn Coursen's citation. Once more we observed full military etiquette, with the crew at attention and Coursen marching briskly to the front and center to receive his award.

This was not just play-acting. The ceremony conveyed a vital message to all of us. We were still members of the U.S. Navy, working as a team within a formal military structure, despite the present circumstances of detention. And we were still on our mission, which we would eventually complete. Military discipline and faith in each other were the strongest tools we had to accomplish that mission. We would continue to perform physical training in our rooms, even though we could not do so as a group. We would continue to say prayers and sing hymns before sitting down to eat at each meal. And next week at this time if we were still here, we would have another awards ceremony. That was all part of the military life we had chosen. When we left the dining hall that night, I sensed a renewed spirit of optimism among the crew.

The next morning after breakfast—the unvarying scoops of rice, steamed vegetables, and a few tasteless shreds of meat—the Chinese surprised everyone by distributing the e-mails that General Sealock had assured us he had turned over to them. Many of them were more than three days old, and I knew the Chinese had withheld them just to harass us.

Although the messages were brief, reading them was like looking through a window that opened directly on our loved ones in America. The e-mail addressed to me was from my dad. "Shane," he wrote, "you'll be proud to know that there are yellow ribbons all over the states of South Dakota and Nebraska. Everyone here is thinking of you and your crew."

We decided to read our e-mails out loud to the group in order to share the warmth and optimism the messages contained. One of the most heartwarming was from the father of Petty Officer Rodney "Ra-Ra" Young of Katy, Texas. His dad got right to the point: "You'd better come home because you promised to help me put up a fence, and I could really use that help." Everyone laughed because the words took us back to the normal world and out of the anxious monotony of our detention.

Immediately after that breakfast, the guards took me for interrogation with the admiral and the base commander. They again had the two model airplanes set up in the glare of a video-camera lamp. Any representation of the collision I might try to make using these models would be deceptive. And I certainly had no idea how the Chinese would edit the videotape. So I refused to even touch the models of the Finback and the Antonov-124.

"How do you feel about our pilot Wang Wei?" the admiral asked, taking a new tack.

"We have not found him yet and fear he is dead," the base commander quickly added.

I had been reading more about Wang Wei in the *China Daily*. "He's a father, a husband, and a son. No one intended a loss of life. It is a horrible thing that he is lost at sea." I meant every word of that because my crew and I had almost been lost that day as well. I would never forget looking *up* through the right windshield at Wang Wei's shattered, flaming aircraft falling below us toward the soft blue surface of the South China Sea.

"Then isn't it right," the admiral asked, "that we demand an apology from yourself and from your government?"

I chose my words carefully. "I am not the investigator, and I am not a government official. I cannot place blame for the collision. To apologize means I accept responsibility for this accident. Therefore I am not willing to apologize."

"You *must* apologize because the collision was your fault," the admiral said.

"No, I cannot."

"We do not feel we are being unreasonable," the admiral said.

"We are not asking for too much," the base commander added.

"If you apologize," the admiral said, "you can leave tomorrow. I know you would like to go home to your families because we also would like to do so."

I nodded in agreement. "We would love to leave tomorrow or right now. But I am *not* willing to give an apology or accept blame for the accident." Again I used the word "accident" rather than "collision" and only hoped that the interpreter would manage to convey the subtle difference. Although it was certainly possible that the admiral himself understood enough English to do so.

He glared at me. "You are being very, very selfish. What about your crew? Do you think they want to stay here? Think about them. You must consider what can happen to them if you stay here, because we cannot allow you to leave until you apologize."

I nodded once more as if accepting his reasoning. Although he might not realize it, the admiral had just offered me the chance for an unscheduled private meeting with the entire crew.

"I would like to discuss this question with all the crew members," I said. "Will we be able to talk freely about this?"

"You can do so," the admiral said, then turned to the guards and issued an order.

Back on the fifth floor, I told the guards and the interpreters there, "The admiral says we are now allowed to speak freely and to move from room to room." What the hell? It was worth a try. After all, when we gave the Chinese an inch on the interrogation rules, they took a mile. We could try doing the same.

Pat and Senior gathered everyone in my room for the meeting. It was pretty crowded in there. Some people sat; others stood around the walls. We had to shut down the roaring air conditioner so that everyone could hear.

"Here's the deal, folks," I told the crew. "I just spent the morning with the admiral and the base commander. You all know me, that I'm straightforward and say what I'm thinking. Sometimes that gets me in trouble. But I don't like to hide things from you. We're all a team."

I looked each crew member in the eye before I continued. They were silent, recognizing I was about to tell them something important.

"They offered us a trip home," I continued. "All we have to do is apologize. It sounds like anyone who does apologize will get to go home. But most of all, they want an apology from the mission commander."

The crew stared at me expectantly.

"The reason I told you this," I said, my voice taking on a hard edge, "is that I will be old and fucking gray before I apologize. You have to think about what *you* want to do. That's up to each of you to decide. But you also have to think about the consequences. I will not return home without my honor. I'll sit in this room until hell freezes over."

Everyone in the crew nodded, their faces set with determination. Obviously, they had already reached their decision. No one was willing to apologize.

"I see that you all agree with me," I said. "I just want to remind you why we're here and whom we represent and that any compromises we make are going to affect us for the rest of our lives. We made it here as a crew together. And we're going to go home as a crew together, with our honor."

Although no one responded, I knew my words reflected their thoughts.

"Folks," I continued, "we've been here nine days, and we might be here quite a few more. But we're doing just fine, and I can't tell you how proud I am of every one of you. I also can't thank you enough for what you've done to help me get through this. Does anyone have anything to say?"

"We want to thank you, sir," Seaman Bradford Borland said.

"You're doing a great job, sir," Senior Mellos added.

Sergeant Richard Pray spoke up. "Sir, we know what you've been going through and what you've been doing for us."

Johnny faced the crew. "We have absolutely nothing to apologize for," he said. "And don't think you'd be helping the crew get home any earlier by apologizing."

Once more, everyone nodded agreement. As I had hoped, we were a united team.

"Hey," I said with a smile, lightening the atmosphere in the room, "after talking to the admiral, the way I interpret the rules is that we can now talk freely and move from room to room."

The crew clapped and hooted.

But Senior Mellos stepped forward and raised his hand. "Listen up," he said. "This really isn't some big breakthrough. Don't get your hopes up too high. We're just back where we were six days ago at Lingshui. We're sure as hell not out of here yet. So don't lose your heads. We need to support each other even more than ever now."

"You've got that right, Senior," I said.

Senior Mellos's words represented Navy leadership at its most effective. Much of what I knew about leadership began in the NROTC program at the University of Nebraska. Although my leadership principles today are much simpler and less formal than those that the guest lecturers on campus proposed, I've found my system works when it has to.

I had always tried to be sincere and straightforward. I'd also learned to really listen when people are talking. They're usually trying to tell you something. If you just pretend to listen, you'll never learn what that is.

When dealing with people both senior and junior to you in rank, talk to them like they're human beings and they'll do the same. Mutual respect is contagious, but it's a trait nearly impossible to fake.

Looking around the crowded room, I was proud that my crew and I had achieved that level of trust and mutual respect.

But when they left the room, I again had to confront the reality of our situation. The Chinese demanded we apologize. Neither my crew nor I would do so. And General Sealock had made it clear our government would never apologize. Suddenly I remembered the accounts of the Iranian hostage crisis that I'd read in college. The Americans at the embassy in Tehran had remained captives 444 days. Would our detention drag on that long? Certainly China would be under international pressure to release us before the decision was made on the host city for the 2008 Olympics and China's admission to the World Trade Organization. But Iran had ignored world opinion and had held our hostages for over a year. I looked around the drab room. This was not a pleasant prospect to consider.

That night at 2030 hours, the crew was called back down to the dining room to meet with General Sealock and Ted Gomm. All afternoon the interrogators had hammered on the officers individually and pairs of enlisted crew members to apologize for the collision. But no one gave in.

Despite the bullying and intimidation, everyone seemed in good spirits. The chance for them to move among the rooms had definitely raised morale.

"Has everyone received an e-mail?" General Sealock asked.

By then, everyone had received at least one. But it was clear the Chinese were still holding some back. The crew nodded and thanked him for delivering the messages.

"General," I said, "we're still doing just fine. And they're still trying to get us to apologize."

"The United States government is not going to apologize," the general said firmly. "What you do is up to you. But again I'm telling you right here that your government will not apologize."

With this clear statement, and following the meeting in my room, I could almost feel the crew's resolve intensify.

We were in this for the long haul. Several crew members wanted Ted Gomm to take powers of attorney for their families in case the detention dragged on for months.

"We can probably hold off on that a day or two," General Sealock said, his tone neutral.

As he turned to leave, I approached him. "General, how are the negotiations really going? Do you have any idea how long we're going to be here?"

He answered very quietly. "I don't know, but you can probably do it standing on your head."

This old military cliché to describe a "short-timer" within days of discharge never sounded so good. General Sealock would not have chosen those words unless he knew that we were nearing the end of detention.

But as I returned to my room, I realized that it was all the more important for us to continue resisting the increasing Chinese pressure to apologize.

Now that we had regained freedom of movement in the corridor, Senior Mellos and I took the opportunity the next day to briefly interview each enlisted crew member individually in my room. We spoke to them about their health, their morale, and asked if they had any concerns back home that we could help with. Then Senior repeated the advice he had given during the meeting the day before.

"Don't screw up. Watch what you say in front of the guards. Stay cool. Remember, we're just where we were seven days ago and we may be here for a while, and we don't want to lose these privileges."

Then I told them what I, as their commander, expected of everyone. "When we do go home, there *will* be no screaming, yelling, no gestures, either here or when we board the plane taking us off this island. I don't want one person's conduct or gesture to demean the rest of us. We need to leave here with the same dignity we had when we arrived. I have complete faith that every one of you will handle yourself professionally." Then I repeated with slow gravity, "Remember, we came as a crew. We'll go home as a crew."

I also met with the commissioned officers to get their opinions on the current situation.

"Do you see any unusual problems?" I asked.

"Some people are nervous about how long the interrogations are going on," Pat said.

Johnny nodded. "The interrogators yelling at them in Chinese is pretty disturbing."

This was important information for me to know. "Look," I said, "these interrogations are a necessary evil for us to get home. Just as long as everyone keeps the faith and doesn't say the wrong thing."

The officers spread out to pass the word. I also visited the rooms of those who the officers had told me were beginning to suffer sagging morale.

That afternoon I again met with the "khakis," Senior Mellos and the officers—so named because our khaki service caps distinguished us from the more junior ranks. "Morale is high right now. But if we're still here through Easter Sunday," I said, "we're all going to have a real hard time dealing with it. I know I will."

They agreed.

"We have to start to prepare for the worst and keep the faith," I told them.

Easter was on April 15, only five days away.

The next morning after breakfast, the guards came to order the entire crew down for a meeting in the dining hall. It was Wednesday, April 11, the morning of our eleventh day of detention. Shuffling down the stairs, many of us speculated that General Sealock must have come for an unusual daytime meeting to announce our release.

Instead, we encountered the somber base commander and a man the interpreter said was a "legal expert." Thc expert launched into yet another propaganda session.

He read from a detailed document, which was no doubt meant to impress us with its authority.

Over and over again, he recited, "You can be tried for . . ."

Ramming the F-8 fighter and killing its pilot.

"You can be tried for . . ."

Violation of Chinese airspace.

"You can be tried for . . ."

Violation of sovereign Chinese territory.

"You can be tried for . . ."

Espionage against the Chinese people.

Although the Chinese wanted us to believe this was a list of formal charges, it was, in effect, just a paraphrase of material we already had read in the *China Daily*. No one believed this diatribe, of course, but we had to sit there and listen.

Then the base commander rose and read from another document.

"Your secretary of state, Mr. Colin Powell, has made a full apology to the Chinese people," he said. "We expect the same from you. You must do this before you can go home."

I glanced around the room. Although no one revealed strong emotion, I knew my crew well enough to recognize they took this assertion exactly for what it was, bullshit. Besides, all of us had read Secretary Powell's diplomatically phrased expressions of "regret" in the *China Daily*. His statement was a long way from a full apology.

Another thought suddenly occurred to me. Just supposing Colin Powell *had* apologized on behalf of the U.S. government. Why would they need another apology from the crew of an airplane commanded by a lowly Navy lieutenant?

They were getting desperate, and we knew it.

After lunch, the base commander came to my room.

"Can I have your identity card?" he asked. "We need to take a picture of it."

That got the alarm bells ringing. *What the hell is going on?* I didn't want to surrender my ID card unless I absolutely had to. "No," I said. "It is my only source of identification."

To my surprise, the base commander did not insist, as he would have in the past. "All right," he said. "We will take pictures of five people at a time. Then we will photograph their identity cards and give them back immediately. You can stay with them."

When we assembled in the dining hall for the photo session, I realized the pictures might have something to do with the Chinese immigration bureaucracy. We certainly had not carried passports with us when we entered China. So maybe these pictures of our ID cards were going to have to serve instead for us to be issued exit visas. I certainly hoped so.

The picture-taking was improvised at best. Because there was not enough light in the room, the photographer set up a chair on the adjacent roof to catch the afternoon sun, placed the cards one by one on the chair, and then snapped a close-up of each of them.

As the pictures were taken, the mood in the room became increasingly relaxed. The base commander was smiling now and chuckling with pleasure. He came over and stood beside me. "Lieutenant Osborn, you are a good officer and care for your people. I want a picture with you for a souvenir."

I was wary of the propaganda value the picture might provide the Chinese. So I stood at attention, my face blank, while the photographer snapped. But the base commander was grinning warmly.

And I was amazed when several of the other officers, including Flattop and the interrogation interpreters, also requested pictures with me. They were relaxed and smiling for the first time since I'd seen them.

*This is it*, I thought, *the end. We're going home.*

Dinner that night was both tense with anticipation and very cheerful because the Chinese brought in a big stack of e-mails and Web site news on American sports that they had been holding back.

But we were marched back up to our rooms with no news that the release we all hoped for was imminent.

Just before midnight, the guards again assembled us and we were ordered to proceed to the dining hall. I expected General Sealock would be there for one of his late-night meetings, and knew he would give us an update on our status.

But when we filed into the room, we were surprised to see that the chairs had all been removed, so that we had to stand in ranks. Then the admiral and the Lingshui base commander entered followed by several other officers. For the first time, they all wore uniforms. In contrast to the almost jovial atmosphere of the afternoon, the Chinese officers' faces were grave.

The admiral began to read from yet another long document. Once more, we were accused of four major breaches of international and Chinese law: espionage against the Chinese people; violating Chinese airspace; ramming the F-8 fighter, resulting in its pilot's death; and violating sovereign Chinese territory by landing at Lingshui without permission. As usual, this list of charges was presented with a detailed recitation of the appropriate clauses and subparagraphs of United Nations conventions and Chinese criminal law.

As the admiral droned on, we shifted wearily from foot to foot. It certainly did not seem as if we were on the verge of release. Looking around the room, I noted the grim but determined faces of the crew. I knew that—even if they marched us back upstairs—our resolve never to cave in and apologize would remain firm.

Finally, the admiral lay down the document and lifted a single page. "But in the interest of humanitarianism," he said, "you are now being released. Transportation is being arranged as we speak."

No one in the crew moved or made a sound. We stood in silent discipline, none of us betraying our emotions. But I had to hold back a smile. Earlier, the officers had joined a betting pool trying to guess the day and time of our release. I had drawn April 11, before midnight. But I hadn't thought I stood much chance of winning because midnight was almost here. Now I had just picked up one hundred twenty bucks off my fellow khakis. But, following naval tradition, I'd probably spend at least that much standing them—and the rest of the crew—to rounds of drinks once we got home. I wasn't going to mind that at all.

The admiral seemed puzzled by our reaction. "You need to understand that this is not a sign of weakness on our part, but a sign of humanitarianism. Now you must go to your rooms, gather your belongings, and prepare to leave."

As we turned to file out, the admiral held up his hand. "Also," he said, smiling now, "your representatives have provided three cell phones. Each of you can call your home for two minutes."

A guard officer placed the three phones on a table in front of us. The crew looked toward me, and I nodded. Pat Honeck and Johnny Comerford stepped back with me. We would be the last to call. It was best to let the others have their chance in case time ran out. The Chinese were watching us closely, but Senior Mellos had the enlisted crew members quickly lined up by ascending order of rank, and the other officers let the more junior crew members call first.

Finally, Pat and Johnny made their calls, and it was my turn. I was frustrated when my mother's phone was busy after four tries. So I punched in the numbers of my dad's home in South Dakota and got his answering machine. "Shane," the recorded voice said in rapid excitement, "we're on our way to Nebraska, and we'll see you when you get home."

I hadn't used my full two minutes, so I tried my mother once more. This time she answered.

"Where *are* you?" she said.

"Mom," I said, my voice hoarse with fatigue. I felt physically and mentally spent. "I'm in China. But we're coming home."

She seemed momentarily confused. "You're still in China?"

"Don't worry, Mom," I said. "We *are* coming home."

As typical during our detention, there were delays leaving the hotel. We had all collected our meager belongings to save as souvenirs, and most of the crew was roaming around the fifth floor corridor, which was thick with cigarette smoke from the crew and guards. Our faithful interpreters, Lieutenants Gump and Tony, signed goodwill messages in Chinese and English for some of the crew. Pat Honeck even taught them the lyrics to the old rock song, "Hotel California:"

*"You can check out. But you can never leave."*

It wasn't until just before dawn on April 12 that we were assembled in the hotel lobby, and an officer pointed toward a large bus parked at the gate with its engine throbbing. "Go on quickly," he said, "and do not move the curtains."

There were Chinese and foreign television crews crowded on the sidewalk opposite the bus, the lamps of their minicams blazing. A throng of armed policemen surrounded the bus. Each of them seemed to be carrying a small flash camera. And when we climbed aboard the bus, they followed us in small groups and fired away with their cameras. They must have wanted their share of the "I was there" action.

As the bus finally began to roll, I saw through the windshield that the streets were lined by hundreds of police.

We left the city by the same route we had entered. As we drove toward the nearby international airport, farmers pushed bikes laden with bamboo baskets of vegetables or led oxcarts piled with cages of chickens and ducks along the dusty side lanes toward Haikou. Smoke rose in wisps from thatch-roofed houses, and I caught the charcoal smell of cooking fires.

That was timeless China outside the window, I realized through my near stupor. We had been detained here almost twelve days, during which I might have accumulated a total of fifteen hours uninterrupted sleep. I was proud that I had endured. But my proudest accomplishment had been protecting my crew, especially the junior enlisted members, from harsh, protracted interrogation. Everyone had held up well to the accusations and threats. No one had broken and offered an apology. Were returning with our honor intact.

The bus left the freeway at the airport exit. Ahead, I saw the tall tail of a Continental Boeing 737 on the tarmac. My crew were tense, expectant. When we pulled onto the apron and parked near the plane, General Sealock and Ted Gomm, both smiling broadly, climbed on board the bus to brief us.

"They just have to refuel the jet," the general said. "It will be only a few minutes longer."

Ted grinned and explained that the Chinese had insisted on receiving a thirty-four-dollar visa fee for each of us.

Finally, we left the bus and marched toward the plane between a cordon of Chinese secret police guards in cheap suits with radio earpieces. When we reached the foot of the airliner's stairs, I stood aside and counted every member of my crew as they climbed aboard, just as I had for every crew movement throughout our detention.

Then I turned to General Sealock and saluted. "Thank you, sir, for everything you did for us here." We shook hands. "And thank you," I added, "for fighting for those sixty seconds."

The airliner had been set up as a mini-evacuation hospital with a medical staff as well as a mini-debriefing center. The doctors quickly checked us out and we took our seats for takeoff. Fifteen minutes later, we were climbing into the hot tropic sunrise, and I suddenly remembered the sunrise above the vast rim of the Pacific on Sunday, April 1.

"We have now entered international airspace," the Continental captain announced over the PA system.

Now the crew cheered loudly.

The first two people I had spoken to on entering the plane were my squadron skipper, Commander Bernard Lessard, and Commodore Captain Kevin O'Brien, in charge of all P-3 operations in the Pacific. Like the other military personnel on board, they wore civilian clothes.

We were scheduled for a four-hour flight to Guam, where a giant Air Force C-17 transport would carry us on to Hickam Air Force Base in Hawaii for debriefing. The skipper wanted to know if I would like to just sleep for a while, or whether Johnny and I were up to discussing the collision and the detention with them.

Johnny and I were too keyed up to sleep. So we spent the flight reliving the previous eleven days. It was not a pleasant experience.

But our brief stop in Guam definitely was pleasant. The people at Anderson Air Force Base pulled out the stops for us. They even had the high school band playing patriotic tunes that were just as spirited as those I played with my own high school band back in Norfolk, Nebraska, on Veterans Day. In the Bachelor Officers Quarters, the Navy had brand new flight suits laid out for us and had flown down our personal gear from the Habu Hilton in Kadena. The Navy even had a barber standing by. And there was a fantastic buffet, with no rice and no fish heads.

There was also American satellite television and the *International Herald Tribune.* Our release from China was the newspaper's lead story. I learned that the tense negotiations had finally concluded when the Bush administration molded diplomatic language concerning the collision that the Chinese government found acceptable.

While refusing to take responsibility for the accident, the letter that Ambassador Joseph W. Prueher had delivered the Chinese Foreign Ministry the day before employed these words: "Both President Bush and Secretary of State Powell have expressed their sincere regret over your missing pilot and aircraft. Please convey to the Chinese people and to the family of pilot Wang Wei that we are very sorry for their loss." Although this was expression of sorrow at the loss of life, it certainly was not the full apology the Chinese had demanded of us for eleven days on Hainan.

"Although the full picture of what transpired is still unclear," the letter from the ambassador to the Foreign Ministry continued, "according to our information, our severely crippled aircraft made an emergency landing after following international emergency procedures. We are very sorry the entering of China's airspace and the landing did not have verbal clearance, but very pleased the crew landed safely."

Although the Chinese government translated the words "very sorry" to mean an expression of sincere apology that involves an acknowledgment of error and acceptance of responsibility, the original English text obviously did not express this meaning. The letter had merely stated American sorrow at the loss of life, and had not accepted blame for the collision. And by describing our aircraft as "severely crippled" and adding that we had followed international procedures (in making repeated Mayday calls and contacting Lingshui directly to declare an emergency), the letter refuted earlier Chinese contentions.

We had held out. The Bush administration had held firm. The Chinese had blinked. Now we could go home.

I sat in the cockpit of the Air Force C-17, the "Spirit of Bob Hope," for takeoff just after dark. It was an incredible airplane, with a few digital computer screens replacing the banks of instruments in our flight station. The contrast between this airplane and the peasants I had seen at dawn pushing their loads of produce beside the road near Haikou could not have been more striking.

As we turned onto the runway for takeoff, I saw a gray P-3 Orion parked on the apron. Lieutenants Chris Rush and Kevin Queen had their crew lined up before the plane. Everyone held their survival flashlight with its red lens in his hand as they saluted us. Watching this warm tribute, I felt tears on my face.

I was back in the vast cargo bay, drifting off to sleep, when the commodore called me. "Lieutenant Osborn," he said, "they need you up front. The secretary of defense wants to speak to you."

I shook myself awake and went forward to the cockpit. The flight crew gave me a headset, and I heard a voice at the other end of the very clear radio connection asking me to stand by for Secretary of Defense Donald Rumsfeld.

The secretary came on the circuit and I thanked him on behalf of the crew for everything the Defense Department had done for us.

Then the secretary said, "I'd like to ask you just two questions." They concerned issues that so far only the Chinese government had commented on publicly: how the collision had occurred and whether we had requested permission to make an emergency landing in Lingshui.

I had never spoken to a secretary of defense before. And I wasn't sure how much I could discuss on this open radio channel. So I turned to Commodore O'Brien. "Sir, can we talk about the flight to Mr. Rumsfeld?"

He smiled. "The secretary of defense just asked you to do so, Lieutenant."

*All right.* It felt great to provide these details. "The only turn the aircraft took was after impact," I explained. "We had a severe left roll after impact into my number one propeller and nose. . . . We were using autopilot, steady altitude and course throughout the entire event until impact. . . . We issued about ten to fifteen guard calls and were unable to hear any response due to holes in my pressure bulkhead." I described flying the circle offshore and that we did not cross over the beach until we had the field in sight. Nor did we set up for the landing until we had overflown the runway to make sure it was clear.

"I'm glad to hear that firsthand," Secretary Rumsfeld said. "And I hope you will give my great respect to your crew."

Perhaps the highlight of our two days at Hickam Air Force Base near Honolulu was the call from President Bush. We'd been in a conference room, on a break from the intense debriefings, when a speaker announced, "Stand by for the President of the United States."

He told the crew that we had done a "great job."

Then he spoke directly to me. "Shane, as an old F-102 pilot, you did a heck of a job bringing that plane down. You made our country proud."

"On behalf of the crew, sir," I told the president, "I thank you for getting us home so quickly."

After two days of the exhausting debriefings, we held a brief news conference beside the Navy C-9 jet transport that would carry us home to Whidbey Island. "I want to thank America, the administration, and everyone involved in getting us home so quickly," I told the reporters, and then I turned away from the microphones to face the crew members lined up behind me. "We can all be proud of this crew."

Directly addressing the Chinese claims that we had rammed the F-8 fighter, I emphasized that our aircraft had been flying "straight, steady, holding altitude, on autopilot, and heading away from Hainan Island" when the Finback had impacted our prop and nose, sending us into a snap roll and near-inverted dive.

Then Senior Mellos commented on the teamwork that had saved our plane from almost certain disaster. "Thank God for the training we practice every day," he said. "Because without it, we'd have a different kind of press conference today."

A reporter asked me about the persistent Chinese demands that our government apologize for the collision. No apology was necessary, I said, because we had not been at fault. "I'm here to tell you that we did it right."

Once the C-9 took off, I could focus on the reunion that lay ahead. On Guam and in Hawaii, I had been able to make long calls to my family and to Roxanne. But talking on the telephone and actually being with the people you loved was a different matter. And I would be with them again soon.

The Navy had done a great job of flying our immediate families to attend the big reunion party at Whidbey Island. And Ross Perot and other patriots had provided free tickets and rental cars to more than a hundred of our other loved ones who were not officially in our immediate families.

As our jet rolled to a stop at the VQ-1 hangar, we looked out at an immense crowd waving American flags and yellow ribbons. The hangar itself was decorated with a huge flag. When the door opened and we climbed down the ladder, a Navy band played "Anchors Aweigh."

Our families rushed forward to greet us. I was so disoriented and exhausted that I didn't even see Roxanne until she threw herself against my chest, almost knocking me over. As I opened my arms, Roxanne pulled close. Moments later, Mom embraced me. Then Dad and his wife, Julie, wrapped me in their arms. I reassured everyone, "I'm fine, I'm fine."

Up on the VIP platform with the admirals and the political officials, I longed for the ceremony to end so that I could be with Roxanne and my family. So when I was asked to speak, I kept my comments brief.

"I'd like to thank God for allowing my crew and myself to be here today," I said. "It was definitely Him flying that plane."

The audience broke into applause, and my crew rose to make this a standing ovation.

I gazed at them proudly. "I'd also like to thank my twenty-three other crewmates." I looked at Johnny, Senior Mellos, Pat and Jeff, Wendy and Regina. Seaman Brad Borland grinned up at me. Petty Officers Shawn Coursen and Ra-Ra Young beamed with pride. "Without them," I continued, "I wouldn't be standing here right now. They performed far, far beyond the call of duty."

Now, one by one, the crew stopped clapping and stood silently to accept their country's gratitude.

We were home.

Epilogue

---

# WASHINGTON, D.C.
# THE WHITE HOUSE

18 May 2001, 0840 Hours

President George W. Bush was true to his word. When he had called the crew at Hickam Air Force Base, the president had asked us to come visit him here at the White House. Now we stood in the Oval Office in a loose crescent formation, wearing dress white uniforms. The president faced us from in front of his desk in the center of the room. Vice President Dick Cheney smiled toward the crew from the left, while National Security Adviser Condoleezza Rice watched from the right.

The president recalled his White House invitation during the debriefing in Hawaii. "The vice president and I are thrilled to be able to look you in the eye and say, 'Thanks for your service to the country.'"

I glanced to my right down the line of my crew. Once more their eyes shone with pride.

"You handled yourselves with such class and dignity," President Bush said, explaining why he had asked us here. "It was important for our nation to realize the fine caliber of people that serve our country. And we're really proud of you. We appreciate your mission; but most of all we appreciate your character. And so it's my honor to welcome you here. I look forward to giving you a tour around this majestic office, a shrine to the greatness of America."

The president then proceeded to mingle with the crew, shaking hands and chatting in a friendly and natural manner. After about fifteen minutes, he smiled and looked at his watch. "'Fraid I've got things to do," he said, sweeping his hand around the office. "But you can stay here and take some pictures if you'd like. You've certainly earned the right to do so."

After the president left, we tentatively wandered around the historic, beautifully decorated office. Then, one by one, the cameras' strobe cubes began to flash.

Then Ra-Ra Young asked an aide if we could take a picture behind the president's desk.

"Sure," the man said. "The president said you could."

"Sir," Senior Mellos said, pointing to President Bush's wide leather chair, "you take a seat right there, and we'll have the flight station stand behind you."

I sat down gingerly, and the flash popped.

We marched up to the grandstand at Andrews Air Force Base at eleven that morning. Here the crew would receive medals for our performance in the air and on Hainan Island. I would be honored with the Distinguished Flying Cross, the armed services' highest decoration for airmanship. Senior Mellos and I would also receive the Meritorious Service Medal for the leadership we had shown during the detention on Hainan. Senior and all the other members of the crew would also receive the Air Medal, recognition that they had performed skillfully and bravely in an extremely hazardous operational mission.

We lined up in our straight formation, waiting for the citations to be read and the Chairman of the Joint Chiefs of Staff, General Henry H. Shelton, accompanied by Secretary of Defense Donald H. Rumsfeld, to present the decorations.

I had always been awed by the Distinguished Flying Cross. I wished there were some way I could share it with the rest of my crew. But that was not possible. Instead, I would accept it in their name and in the spirit of their courageous service.

"The President of the United States takes pleasure in presenting the Distinguished Flying Cross to Lieutenant Shane J. Osborn, United States Navy . . ." The amplified voice of the master of ceremonies carried across the concrete parking ramp toward the EP-3E that we had flown from Whidbey Island the day before. "For extraordinary achievement while participating in aerial flight . . . during sensitive surveillance operations on 1 April 2001 . . . following an in-flight collision with a People's Republic of China fighter aircraft, Lieutenant Osborn displayed superb airmanship and courage. Despite extreme damage to the aircraft . . . he heroically regained control, directed appropriate emergency procedures, and coordinated the crew's efforts to safely land the aircraft. Lieutenant Osborn's dedicated efforts ensured the survival of twenty-four crew members. . . . By his superb airmanship, proven ability to perform under pressure, and steadfast devotion to duty, Lieutenant Osborn reflected great credit upon himself and upheld the highest traditions of the United States Naval Service."

I came to attention and saluted. General Shelton returned my salute and pinned the two decorations to the chest of my uniform. This was a very proud moment. I suddenly remembered those sunny afternoons in Leslie Brewer's little Piper Cub over the Dakota cornfields. I recalled the hours of studying basic aeronautics in Civil Air Patrol in Norfolk. The drudgery of Aviation Preflight Indoctrination came back, as did the tension of Lieutenant Jeff Nelson's stern flight instruction in Primary and the joy of my first solo flight. The hours and hours of emergency procedure training in C-12s flashed in vivid sequence through my mind. Then there were the days and weeks in the cockpit of the EP-3E, during which my most demanding mentor, Lieutenant Norm Maxim, had shaped me into a mature airman and a leader in his own model.

But the man standing at my left, Senior Chief Aviation Mate Nicholas Mellos, had to rank high among those who had brought me here today. I was grateful that he was my colleague and my friend.

General Shelton and I exchanged another crisp salute. Still at attention, I glanced down the line of the crew.

I was proud beyond words of them all.

## Afterword

For over a month after our safe return from Haikou, the Chinese and American governments negotiated the fate of our crippled EP-3E stranded at the Lingshui Air Base. The Americans wanted to repair the plane and fly it out of China. The Chinese would not accept this proposal.

Finally, in late May, the two sides reached a compromise. The United States would dispatch a team of civilian technicians from the Lockheed Martin Aeronautics Company, who would disassemble the aircraft so that it could be flown out of Hainan in pieces aboard two giant Antonov-124 cargo planes leased from a Russian firm.

In mid-June, the Lockheed-Martin team arrived at Lingshui to begin work. They drained the aircraft's fuel tanks, oil, and hydraulic fluid, and then packed up small equipment in crates. Next they removed the big "M&M" antenna cover, and smaller covers and antennas. The crew then removed the engines and propellers and the tall tail assembly. Finally, the wings came off, the long lines of rivets sheared with power tools.

On July 4, 2001, the big Antonovs began shuttling their loads between Lingshui and Kadena Air Base, Okinawa, where our flight had begun. A week later, the Antonovs landed at Dobbins Air Force Base near the Lockheed-Martin factory in Marietta, Georgia.

The EP-3E is still being analyzed to determine if it can be safely rebuilt to fly again.

Meanwhile, the Chinese submitted a bill for one million dollars to the U.S. government to cover expenses associated with our detention on Hainan and the recovery of the aircraft. The American Embassy in Beijing called the bill "highly exaggerated." And the U.S. House of Representatives voted 424 to 6 against giving the Chinese any compensation at all for expenses related to our detention or the return of the EP-3E surveillance plane.

The issue remains unresolved.

## Glossary of Aviation Terms

*Aerobatic flight*
Maneuvers in which an aircraft departs from normal flight attitudes to perform rolls, loops, and dives, while remaining under control.

*Aileron*
One of an aircraft's movable control surfaces, located on the outer, trailing edge of each wing, which controls the angle of bank or roll.

*Airframe*
The basic structure of an aircraft: the fuselage, wings, tail, etc., excluding the engines, instruments, and nonstructural internal equipment.

*Airspeed*
The speed of an aircraft through the surrounding air, which, because of the effect of winds, does not necessarily reflect the aircraft's speed across the ground below (groundspeed).

*Angle of Attack (AOA)*
The angular distance of an aircraft's lifting surfaces (wings, etc.) above the horizontal. Too high an Angle of Attack at too low an airspeed will result in a stall.

*Attitude*
The position of an aircraft in flight relative to the earth, i.e., wings level, nose high, nose down, inverted, etc.

*Bank*

Maneuver controlled by the ailerons in which an aircraft dips one wing while raising the other; the basic method of changing horizontal direction. In straight and level flight, an aircraft has a zero-degree angle of bank; fully inverted, the bank angle is 180 degrees. Aircraft are limited by design from exceeding certain angles of bank.

*Cockpit (Flight Station or Flight Deck)*

The position at the front on an aircraft from which pilot(s) operate the controls. On older, multiengine aircraft, a flight engineer may also work in the cockpit.

*Control surfaces*

Movable sections on the wings and tail—ailerons, elevators, and rudder—that permit an aircraft to maneuver.

*Drag*

The force holding back an aircraft in flight from forward or vertical movement. Modern aircraft are designed to minimize drag.

*Elevator*

Movable control surfaces on the trailing edge of an aircraft's horizontal tail that govern the Angle of Attack.

*Feather (Propellers)*

To manipulate the blades of a propeller of a shut-down engine so that they are edge-on to the airstream in order to minimize drag.

*Firewall*

Aviator's jargon for the emergency use of maximum engine power exceeding Military Power.

*Flaps*

A pair of matched extendable panels on the wings' trailing edges. At takeoff and landing, and at slower speeds, flaps increase the wings' surface and thus increase lift, but also increase drag. Flaps are used to slow the aircraft for landing.

*Fuselage*

The tubular or cylindrical body of an aircraft to which the wings and tail are attached.

*G forces (Gs)*

The force an aircraft and its occupants experience during flight maneuvers. One G equals the normal force of gravity; Gs can be either positive or negative.

*Groundspeed*

An aircraft's true speed across the ground, which, due to the effect of winds, might differ from airspeed.

*HF*

High Frequency radio, used for long-distance communication.

*Hydraulic system*

An aircraft's "power steering," which uses hydraulic fluid under pressure to move control surfaces.

*Leading edge*

The forward edge of the wing or the tail's horizontal stabilizer.

*Lift*

The lifting force generated by an aircraft's aerodynamic surfaces, especially in the wings and tail.

*Mayday*
International emergency distress radio call, which an aircraft usually makes over the Guard Channel frequency of 121.5 or 243 MgHz.

*Military Power*
The maximum nonemergency power setting for the engines of military propeller-driven aircraft.

*"Mushy" flight*
Aviator's jargon for unstable flight, usually resulting from flying too slowly at too high an Angle of Attack.

*Nose cone*
A streamlined cone at the front of an aircraft, usually covering a radar antenna.

*Pitch*
The angle of an aircraft's nose relative to the horizon, which is controlled by the elevators.

*Pitot tube*
A hollow tube mounted on an aircraft's fuselage that feeds air to an instrument, especially the airspeed indicator.

*Power lever*
Also called the throttle on many aircraft: the engine control that governs the amount of fuel an engine receives (especially turbine engines).

*Propeller pitch*
The angle or "bite" of propeller blades relative to the airstream.

*Radome (Radar dome)*
A streamlined covering or fairing for a radar antenna.

*Roll*

An aircraft's tilting movement (bank) during which one wing dips and the other rises; rolls can accompany dives and climbs, especially during aerobatic or uncontrolled flight.

*Rudder*

The large control surface in the tail's vertical stabilizer, which governs an aircraft's sideways movement. The pilot controls the rudder by pushing the right or left rudder pedal.

*Snap Roll*

Sudden, extreme roll maneuver. A snap roll will usually exceed the safe angle of bank of all aircraft except those designed for aerobatic flight.

*Stabilizer*

One of the T-shaped surfaces of an aircraft's tail; there are usually one vertical and two horizontal stabilizers.

*Stall*

Loss of lift with resulting uncontrollability that occurs when an aircraft flies either at too high an Angle of Attack or too slowly, or at a combination of both.

*Stick*

Sometimes called control stick: The device used principally in smaller aircraft to control roll and pitch. Sideways movement of the stick activates the ailerons; pushing and pulling the stick activates the elevators.

*Trailing edge*

The aft (rear) edge of a wing or horizontal stabilizer.

*Turboprop*

An aircraft engine that combines a jet turbine with a propeller.

*UHF*

Ultra-High Frequency radio, used for short-range communication (within line of sight), which depends on an aircraft's altitude above receiving station.

*Vertical Speed Indicator (VSI)*

An instrument that measures in feet per minute an aircraft's rate of climb or descent.

*Windmill*

An in-flight emergency in which the propeller of a shut-down engine fails to feather and continues to spin.

*Yaw*

The flat, sideways turn of an aircraft's nose to the left or right.

*Yoke*

Sometimes called the control yoke: The device similar to a steering wheel (cut in half), used principally in larger aircraft to control roll and pitch. Sideways movement of the yoke activates the ailerons; pushing and pulling the yoke activates the elevators.